REALIZING THE
PROMISE OF
PRECISION MEDICINE

REALIZING THE PROMISE OF PRECISION MEDICINE

The Role of Patient Data, Mobile Technology, and Consumer Engagement

PAUL CERRATO

JOHN HALAMKA

ACADEMIC PRESS

An imprint of Elsevier

Academic Press is an imprint of Elsevier
125 London Wall, London EC2Y 5AS, United Kingdom
525 B Street, Suite 1800, San Diego, CA 92101-4495, United States
50 Hampshire Street, 5th Floor, Cambridge, MA 02139, United States
The Boulevard, Langford Lane, Kidlington, Oxford OX5 1GB, United Kingdom

Notices

Knowledge and best practice in this field are constantly changing. As new research and experience
broaden our understanding, changes in research methods, professional practices, or medical treatment
may become necessary.

Practitioners and researchers must always rely on their own experience and knowledge in
evaluating and using any information, methods, compounds, or experiments described herein.
In using such information or methods they should be mindful of their own safety and the safety
of others, including parties for whom they have a professional responsibility.

To the fullest extent of the law, neither the Publisher nor the authors, contributors, or editors, assume
any liability for any injury and/or damage to persons or property as a matter of products liability,
negligence or otherwise, or from any use or operation of any methods, products, instructions, or ideas
contained in the material herein.

Library of Congress Cataloging-in-Publication Data
A catalog record for this book is available from the Library of Congress

British Library Cataloguing-in-Publication Data
A catalogue record for this book is available from the British Library

ISBN: 978-0-12-811635-7

For information on all Academic Press publications visit our website at
https://www.elsevier.com/books-and-journals

Working together
to grow libraries in
developing countries

www.elsevier.com • www.bookaid.org

Publisher: Mica Haley
Acquisition Editor: Rafael Teixeira
Editorial Project Manager: Tracy Tufaga
Production Manager: Kiruthika Govindaraju
Designer: Mark Rogers

Typeset by TNQ Books and Journals

Disclaimer: The information in this book should not be regarded as legal advice but as educational
content only. Readers should consult their legal or other professional advisors before deciding how to
apply this information in the work place. Similarly, any mention of commercial entities should not be
regarded as endorsements by the authors but is provided for educational purposes only.

DEDICATION

There is honor, love, and family, and work left to be done.

Livingston Taylor

CONTENTS

ABOUT THE AUTHORS

Paul Cerrato has more than 30 years of experience working in healthcare and has written extensively on clinical medicine, electronic health records, protected health information security, practice management, and clinical decision support. He has served as Editor of *Information Week Healthcare*, Executive Editor of *Contemporary OB/GYN*, Senior Editor *RN* Journal, and contributing writer/editor for the  Yale University School of Medicine, the American Academy of Pediatrics, Information Week, Medscape, Healthcare Finance News, IMedicalapps.com, and Medpage Today. The Healthcare Information and Management Systems Society has listed Mr. Cerrato as one of the most influential columnists in healthcare IT. Mr. Cerrato has won numerous editorial awards, including a Gold Award from the American Society of Healthcare Publications Editors and the Jesse H. Neal Award for Editorial Excellence, which is considered the Pulitzer Prize of specialized journalism.

John Halamka, MD, MS, is the International Healthcare Innovation Professor at Harvard Medical School, Chief Information Officer of the Beth Israel Deaconess System, and a practicing emergency physician. He strives to improve healthcare quality, safety, and efficiency for patients, providers, and payers throughout the world using information technology. He has written five books, several hundred articles, and the popular Geekdoctor blog.

Dr. Halamka also serves on one of the advisory committees for the Precision Medicine Initiative, which has been funded with $215 million from the US government.

PROLOGUE: NAKED EMPERORS AND SUPERCOMPUTERS

The Pulitzer Prize winning author and physician Siddhartha Mukherjee has called cancer "The Emperor of All Maladies." [1] And for good reason, it commands the attention of all who come in contact with it, whether they be patients, clinicians, or technologists, claiming the lives of more than 8 million worldwide and disrupting the lives of millions more. But if the promise of precision medicine is fully realized, we may one day come to view an Emperor who has lost his clothes, its "reign of terror" forfeit to searchable medical databases, sophisticated analytics, and an engaged populace.

No doubt many readers will view this vision with a skepticism hardened by years of experience watching "revolutionary" medical advances come and go. Kathy Halamka is not one of them.

Kathy, John Halamka's wife, was diagnosed with Stage III breast cancer in 2011, at which point a sentinel node biopsy revealed that the tumor had already spread to a few nearby lymph nodes. The malignancy was estrogen and progesterone positive but HER-2 negative, less than 5 cm in diameter, poorly differentiated, and fast growing. On average, the 5-year relative survival rate for women like Kathy is 72%, which means that people who have the cancer are only about 72% as likely as people who do not have it to live for at least 5 years after being diagnosed [2].

The standard of care for cases like this is typically chemotherapy followed by mastectomy [3]. But having access to digital resources such as the Shared Health Research Information Network (SHRINE), Informatics for Integrating Biology and the Bedside (i2b2), and Clinical Query 2 presented new options for Kathy and an opportunity to test drive the personalized medicine approach to healthcare.

i2b2 is an open source software platform that gives clinicians and researchers Web-based access to a hospital's electronic health records (EHRs), a resource that has the potential to locate treatment options not yet available in the current medical literature or officially endorsed practice guidelines. You might think of i2b2 as an operating system on which applications such as Clinical Query 2 sit, giving it used friendly access to patient records.

SHRINE is a network of computer systems that are affiliated with Harvard Medical School, giving users access to the EHRs of all of its affiliated hospitals, including Massachusetts General Hospital, Brigham and Women's Hospital, and Dana-Farber Cancer Institute.

We will provide more details on i2b2, SHRINE, and Clinical Query 2 in Chapter 6, but for the purposes on Kathy's narrative, it's enough to know that these were the sources used to individualize treatment of her Stage IIIa breast cancer. When Kathy's providers accessed i2b2, they queried it about a 50-year old Asian female with Stage III breast cancer and asked how many patients seen in all the Harvard-affiliated hospitals fit her profile. The system found over 17,000 and provided the medications they received, their average white blood cell counts, their prognosis, and so on.

The query revealed that this stage of breast cancer was commonly treated with a combination of doxorubicin (Adriamycin), cyclophosphamide (Cytoxan), and paclitaxel (Taxol). But the database search also revealed that many of these patients developed neuropathy—numbness of the hands and feet—from Taxol.[1] Further investigation found that there was only one clinical trial looking at the use of Taxol in this context, and it used a specific number of mg/kg body weight administered in nine doses. There were no data to indicate that this was the optimal dosage regimen or if 3 doses or 11 doses would have resulted in better outcomes, both in terms of tumor shrinkage and adverse effects.

With these findings in hand, Kathy's oncology team decided to personalize her treatment by administering full protocols of Adriamycin and Cytoxan but only a half protocol of Taxol, giving her five doses rather than nine. The individualized approach caused her tumor to melt away and resulted in minimal numbness in her hands and feet—an important benefit considering she is a visual artist who relies on her fine motor skills.

Few patients currently have the resources that Kathy Halamka had. But that is gradually changing. The goal of the precision medicine movement is to give clinicians and patients access to the kinds of information needed to create individually tailored programs to treat a variety of diseases and to ward off those that are preventable. To accomplish those twin goals will require the collection of far more data than clinicians now collect when they evaluate patients. It will require more sophisticated analytic tools to

---

[1] Since this book will serve two main audiences, namely clinicians and technologists, and since they do not share the same specialized language, it will be necessary at times to provide explanations for terms and concepts that may seem obvious to one audience or the other. When possible, this will be done by providing plain English explanations in sidebars or boxes so that readers familiar with the information will not be slowed down as they move through the main text.

glean meaningful insights from the data collected. And equally important, it will require the public to become more engaged in its own care.

This three-pronged approach to personalized healthcare serves as the foundation for the book you are about to read. We will explore the difference between population, one-size-fits all medicine and precision medicine, and take a closer look at the $215 million federally sponsored Precision Medicine Initiative. And since each person's genetic makeup plays a critical role in his/her susceptibility to disease, we will devote a chapter to genomics. There are also chapters on the role of data analytics, EHRs, and mobile technology in delivering personalized healthcare.

Of course, any serious discussion of precision medicine also has to address obstacles and limitations, including the challenge of interoperability and protecting patient privacy, which we have covered in Chapters 8 and 9.

Precision medicine may not be the revolution that some enthusiasts believe it to be, but as the following chapters will demonstrate, it is poised to profoundly transform patient care and consumer self-care by enlisting the technological tools that sci-fi fans only dreamt of a few short years ago.

REFERENCES

[1] Mukherjee S. The emperor of all maladies: a biography of cancer. New York: Scribner; 2011.
[2] American Cancer Society, Breast cancer survival rates, by stage. http://www.cancer.org/cancer/breastcancer/detailedguide/breast-cancer-survival-by-stage.
[3] Strickland E. One woman's fight against cancer in the new era of precision medicine. In: IEEE spectrum. May 28, 2015. http://spectrum.ieee.org/biomedical/diagnostics/big-data-beats-cancer.

CHAPTER ONE

Population Medicine Versus Personalized Medicine

Clinicians and technologists who open this book for the first time may be somewhat skeptical about its premise. Why do we need a new approach to health care? While precision or personalized medicine may be the latest buzz word, the current medical model has served us well for more than 100 years. Why abandon it?

There is no doubt that the era of scientific medicine, with its reliance on laboratory, animal and clinical experimentation, randomized controlled trials, epidemiology, and the astute observational skills of thousands of physicians and nurses has produced profound benefits. The list is too long to enumerate, but even a cursory survey of the history of medicine would have to include insulin therapy, vaccines, X-rays, surgical anesthetics, antibiotics, organ transplantation, oral contraceptives, and cancer chemotherapy. But despite these and countless other advances, paging through any general medicine textbook, one cannot help but notice a troubling consistency. The words "etiology unknown" appear over and over again, revealing the fact that while many disorders are manageable, they are by no means curable. And in far too many cases, the management approach also results in many adverse effects and many "nonresponders." Clearly the adage about the "cure being worse than the disease" is not always the unreasonable compliant of a public that expects too much.

We obscure this lack of a definitive etiology with cryptic terms such as "idiopathic" and "essential"—as in idiopathic scoliosis or essential hypertension—but the implication remains the same. We may have a partial understanding of the pathogenesis of such diseases and have plotted out a detailed pathophysiology in many cases, but root causes remain a mystery.

CURE VERSUS MANAGEMENT

Let's discuss rheumatoid arthritis (RA) as an example. The underlying genetic and environmental events that start the disease process remain

Realizing the Promise of Precision Medicine
ISBN 978-0-12-811635-7
http://dx.doi.org/10.1016/B978-0-12-811635-7.00001-4

unknown, leading eventually to secretion of proinflammatory chemokines and cytokines, activation of T cells and macrophages, and production of several interleukins, tumor necrosis factor-α (TNF-α), insulin–like growth factor, and other agents that bring about deterioration of joints and debilitating symptoms. TNF-α clearly plays a pivotal role in the pathogenesis of the disease, but it is only an *intermediate* cause. The best we can say about the root cause of RA is that like other autoimmune diseases, it probably occurs when a genetically susceptible host is exposed to an environmental antigen.

The fact that TNF-α is part of the pathway from the initial unknown genetic/environmental root cause or causes to full-blown clinical disease has made it a target for pharmacologic agents that inhibit its production. These TNF inhibitors have had a major effect on signs and symptoms, but as Fig. 1.1A illustrates, these drugs also interfere with several biochemical pathways that help the body maintain normal immune functioning. The result of this disruption has proven devastating for many patients.

A recent review of the medical literature sums up the issues this way:

Animal experiments have demonstrated the importance of TNF-alpha in protection against several pathogens including Mycobacterium tuberculosis, M. avium, M. bovis, Bacillus Calmette-Guérin (BCG), Aspergillus fumigatus, Histoplasma

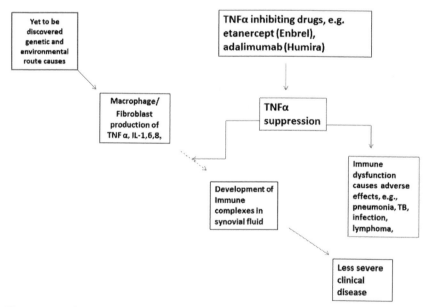

Figure 1.1A Rheumatoid arthritis: etiology, pathology, and management. *IL*, interleukin; *TNF*, tumor necrosis factor.

capsulatum, Toxoplasma gondii, Cryptococcus neoformans, and Candida albicans. These organisms are not killed readily by host defense mechanisms but rather are sequestered within granulomas, which are comprised of a central core of macrophages, multinucleated giant cells, and necrotic debris surrounded by macrophages and lymphocytes. TNF-alpha is required for the orderly recruitment of these cells and for continued maintenance of the granuloma structure.

TNF-alpha is an important component of the immune system's response to a variety of infections, and use of TNF-alpha inhibitors has been associated with an increased risk of serious infections. These include bacterial infections (particularly pneumonia), zoster, tuberculosis, and opportunistic infections [1].

This scenario is true for numerous degenerative diseases and stands in stark contrast to disorders in which root causes are understood, and prevention and complete cures are possible. Iron (Fe) deficiency anemia is a case in point. To keep this comparison simple, consider Fe deficiency anemia induced by a diet lacking in the nutrient. As Fig. 1.1B shows, the result is abnormal red blood cell production, a drop in hemoglobin, and the signs and symptoms of microcytic anemia. But in this scenario, giving the patient a diet rich in iron plus iron supplementation cures the disorder by *restoring*

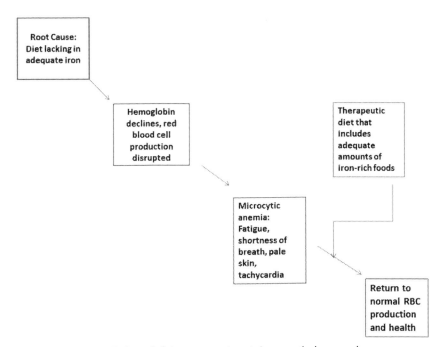

Figure 1.1B Iron deficiency anemia: etiology, pathology, and cure.

a normal biochemical pathway, not by interfering in a pathway that is required for normal body functioning.

Put another way, the treatment of dietary iron deficiency anemia is a form of precise medicine. The treatment of RA with TNF-α inhibitors is not.

THE LIMITATIONS OF POPULATION-BASED MEDICINE

Of course, the reason this anemia can be precisely managed is because it has a single cause. RA, like many degenerative disorders, probably does not. It is this multifactorial nature that highlights the shortcomings of population medicine—the one-size-fits-all approach to patient care. It is also what makes the personalized approach to medicine so promising.

The shortcomings of the one-size-fits-all approach to patient care are well illustrated by the fact that so many patients fail to respond to state-of-the-art therapy for a variety of disorders. By one estimate, the 10 best-selling medications in the United States benefit between 1 in 25 and 1 in 4 patients using them [2]. The current role of statins in the prevention and treatment of cardiovascular disease highlights the limitations of population-based medicine. It has been estimated that only one in 20 patients given rosuvastatin (Crestor) will benefit from the statin [2]. Similarly, one out of every 23 patients will likely benefit from esomeprazole (Nixium) and one in 9 will benefit see an antidepressant effect from duloxetine (Cymbalta).

Research like this does not negate the value of statin therapy in patients with preexisting heart disease, which was confirmed by a 2013 Cochrane Collaboration review, but the same analysis also concluded that there was no net benefit for patients without the disease [3].

A closer look at the research on clopidogrel (Plavix) presents a similar picture. The antiplatelet drug is used extensively in medical practice to reduce the risk of thrombotic events (blood clots). One of the pivotal studies enrolled over 19,000 patients over 3 years and found that the drug was more effective than aspirin in reducing the risk of ischemic stroke, myocardial infarction (MI), or vascular death. Compared to aspirin, it reduced the likelihood of these outcomes by 8.7%, which once again translates into 2 of 100 patients seeing benefit [4].

What is needed, then, is an approach to patient care that identifies those 2 out of 100 individuals who will actually benefit from statins and antiplatelet drugs, as well as all the thousands of other drugs, surgical procedures, and lifestyle modifications recommended by clinicians worldwide.

That is the lofty goal of precision medicine. To accomplish that feat, it will need to take a holistic, personalized view and in many cases, study *massive* population samples to detect significant associations between genetic and environmental risk factors and disease, with the aim of detecting root causes, or in the absence of root causes, identifying which individuals will respond well to available management protocols and which ones will not.

APPLYING THE PRECISION MEDICINE MODEL

Since we began this discussion with the shortcomings of drug therapy for RA, let's use the disorder as an example of the potential advantages of the precision medicine model.

Currently, there is no treatment protocol for the disease that takes into account all the contributing genetic and environmental causes because they remain elusive. Robert H. Shmerling, MD, Clinical Chief, Division of Rheumatology at Beth Israel Deaconess Medical Center explains the issue as follows: "All of our rheumatoid arthritis patients would benefit from a precision medicine approach. We have algorithms to manage RA but they do not take into account much about the patient other than the fact that they need treatment. We cannot predict who will fail methotrexate mono-therapy (which turns out to be 30%-50% of patients), which anti-TNF might work best, whether a biologic other than an anti-TNF might be better than an anti-TNF agent. Many RA patients who went through several medications soon after diagnosis would attest to the aggravation of 'trial and error' and would love to have had personalized/precision/individually tailored treatment up front." So in reality, one-size-fits-all medicine is often trial and error medicine.

What might a personalized approach look like? There are tantalizing hints to suggest what this could be. It has been estimated that genetic factors may be responsible for approximately 60% of one's susceptibility to RA [5]. Several genes have been studied to determine if their presence in an individual is a risk factor for RA. Since RA is an autoimmune disorder, researchers have investigated variations in human leukocyte antigen (HLA) genes, which are involved in the production of proteins that help the immune system distinguish self from nonself. One of the strongest associations exists between RA and the gene HLA-DRB1 [6].

And since RA is an autoimmune disorder, researchers have also been studying the presence of autoantibodies in persons with the disease and those at risk. Two of these antibodies—rheumatoid factor and anticyclic citrullinated peptide—have been correlated with tissue levels of omega-3 fatty acids, which are abundant in fatty fish such as salmon and tuna. More specifically, increased amounts of omega-3 fatty acids in a person's red blood cells have been inversely associated with rheumatoid factor and anticyclic citrullinated peptide positivity, which suggests that this group of nutrients may have a protective effect, reducing the risk of the disease. Even more significant was the fact that the inverse relationship between arthritis and the fatty acids only occurred in persons who were positive for the genetic marker we discussed above. Ryan Gan and his associates point out the following: "The strongest genetic predictor of seropositive RA is the shared epitope (SE) defined by human leucocyte antigen - antigen D related (HLA-DR) alleles" [7].

While research like this has already convinced some clinicians to recommend omega-3 fatty acid-rich foods and supplements, the problem with these and related studies is that they only look at small groups of patients. Gan et al., for instance, only recruited 136 subjects, and it was a case control study, a type of observational investigation that is not considered the gold standard to establish a cause and effect relationship between potential risk factors and a disease.

For a personalized approach to RA or any other disease to convince most thought leaders and clinicians in medicine to change their treatment approach, stronger evidence will need to be presented. The federally sponsored Precision Medicine Initiative (PMI), which is discussed in more detail in Chapter 2, will provide that evidence. It is designed to analyze data on more than 1 million Americans and will involve a major commitment from technologists in a variety of specialties.

HOW DOES THE PRECISION MEDICINE INITIATIVE FIT INTO THE EQUATION?

In 2015, Dr Francis Collins, the director of the US National Institutes of Health, and his PMI Working Group, put together a plan to turn precision medicine's potential into reality [8]. Thanks to advances in genomic technologies, data collection and storage, computational analysis, and mobile health (mHealth) applications over the last decade, the creation of a large-scale precision medicine cohort is now possible. The Working Group identified a number of high-value scientific opportunities that could be used to inform the design of the PMI cohort. These include the following:

- Development of quantitative estimates of risk for a range of diseases by integrating environmental exposures, genetic factors, and gene–environment interactions; identification of determinants of individual variation in efficacy and safety of commonly used therapeutics;

- Discovery of biomarkers that identify people with increased or decreased risk of developing common diseases;
- Use of mHealth technologies to correlate activity, physiologic measures, and environmental exposures with health outcomes;
- Determination of the health impact of heterozygous loss-of-function mutations; development of new disease classifications and relationships;
- Empowerment of participants with data and information to improve their own health; and
- Creation of a platform to enable trials of targeted therapy.

Returning once again to the example of RA, the federal initiative should be able to measure a wide variety of genetic and environmental markers in a massive human database to confirm or refute the risk factors that have been hinted at in small-scale observational investigations.

But what makes this project so noteworthy is the fact that it will not only enlist a million or more volunteers but will also measure an unprecedented number of diverse parameters. Table 1.1, which comes from the PMI Working Group Report, provides a detailed list of what will be measured. In addition to the usual demographic details, including date of birth, gender, race, marital status, educational status, occupation, and income, it will include the following:

- Self-reported measures such as family history, pain scale readings, quality of life assessment
- Lifestyle measures such as dietary intake, physical activity level, use of alternative therapies
- Sensor-based data from smartphones and wearable devices, including respiratory and heart rates and activity level
- Clinical data from electronic health records
- Health-care claims data
- Genomics, proteomics, standard clinical chemistry
- Data from social media feeds
- Environmental data such as exposure to toxic metals, air quality, population density

This nationwide experiment portends to be the largest holistic data collection and analysis initiative in history. An evaluation of this massive database should be able to reveal specific genetic markers for RA, as well as identify dietary and psychosocial risk factors for the disease. It will likely turn up completely unexpected correlations as well. For example, a potential link between intestinal flora and RA has been studied recently. The federally sponsored PMI may be able to identify specific bacterial strains that are causative, or protective.

Table 1.1 Parameters to be measured by the Precision Medicine Initiative (PMI)

Category	Examples	Source(s)	Example uses
Individual demographics and contact information	Date and place of birth, sex and gender, detailed and multiple races/ethnicities (e.g., Asian of Indian descent, Asian of Chinese descent), name, mailing address, phone number, cell phone number, email address, marital status, educational status, occupation/income	Study participant, health-care provider organizations	Participant-specific communications, analytics, risk stratification, assessment of covariates and confounds, study appointment reminders, invitations to participate in substudies
Terms of consent and personal preferences for participation in the project	Fine-grained consent for options to participate, e.g., receive research results	Study participant	"Precision Participant Engagement"
Self-reported measures	Pain scales, disease-specific symptoms, functional capabilities, quality of life and well-being, gender identity, structured family health history	Study participant	Many
Behavioral and lifestyle measures	Diet, physical activity, alternative therapies, smoking, alcohol, assessment of known risk factors (e.g., guns, Illicit drug use)	Study participant (retrospective and prospective) and health-care provider organizations	Correlation with clinical events, drug response, and health outcomes
Sensor-based observations through phones, wearables, home-based devices	Location, activity monitors, cardiac rate and rhythm monitoring, respiratory rate	Smartphone sensors, commercial and research-grade physiologic monitors	Functional ability and impairment assessment
Structured clinical data derived from Electronic Health Records (EHRs)	ICD/CPT billing codes, clinical lab values, medications, problem lists	Multiple provider organizations per study participant, via institutionally managed channels or direct from participant via personal download/upload	Correlation of clinical events with other categories of data
Unstructured and specialized types of clinical data derived from EHRs	Narrative documents, images, EKG and EEG waveform data	Multiple providers, via federated queries rather than inclusion in core dataset	Correlation of clinical events with other categories of data

PMI baseline health exam	Vital signs, medication assessment, past medical history	Study participant interacting with health-care provider organization	Provides baseline measures on all participants
Health-care claims data	Periods of coverage, charges and associated billing codes as received by public and private payers, outpatient pharmacy dispensing (product, dose, amount)	CMS and other federal sources, private insurers, pharmacy benefits management organizations	Assessments requiring complete longitudinal record of exposures/outcomes during specific periods, e.g., within X years of a diagnosis or medication exposure, health services research, exposure and outcomes assessment
Research-specific observations	Research questionnaires, ecological momentary assessments, performance measures (6 min walk test), disease-specific monitors (e.g., glucometers, spirometers)	Study participants, research organizations	Many
Biospecimen-derived laboratory data	Genomics, proteomics, metabolites, cell-free DNA, single cell studies, infectious exposures, standard clinical chemistries, histopathology	Study participants, provider organizations, outsourced laboratories	Correlation of tissue findings and high-throughput biomolecular data with other categories of data
Geospatial and environmental data	Weather, air quality, environmental pollutant levels, food deserts, walkability, population density, climate change	Public and private sources not directly part of PMI	Epidemiology, epidemic surveillance
Other data	Social networking e.g., Twitter feeds, social contacts from cell phone text and voice, OTC medication purchases	Public and private sources not directly part of PMI	Predictive analytics

CMS, Centers for Medicare and Medicaid Services; CPT, current procedural terminology; ICD, International Classification of Diseases; OTC, over the counter. From Precision Medicine Initiative (PMI) Working Group. The precision medicine initiative cohort program – building a research foundation for 21st century medicine. September 17, 2015. http://acd.od.nih.gov/reports/DRAFT-PMI-WG-Report-9-11-2015-508.pdf

PUTTING THE PRECISION MEDICINE MODEL TO WORK NOW

While the precision medicine model currently has little immediate benefit for RA patients, the model can have an immediate impact on the prevention or management of several other disorders. Diabetes is a prime example.

Research conducted by Jeremy Sussman and his colleagues at the University of Michigan and Tufts Medical Center demonstrates that the promise of precision medicine is being realized now. That realization comes in the form of predictive analytics that can help identify individuals at risk of developing type 2 diabetes and, equally important, help identify those who will not develop the disease despite having certain risk factors [9].

Sussman et al. analyzed the results of the Diabetes Prevention Program (DPP), a large-scale randomized clinical trial in which over 3000 patients were given various preventive regimens to determine if it were possible to reduce the incidence of type 2 diabetes in at-risk patients. The participants were divided into three groups:

1. Standard lifestyle recommendations plus twice daily placebo pills,
2. An intensive lifestyle modification program, which included lessons from a case manager on implementing the regimen, and
3. Standard lifestyle recommendations plus 850 mg of metformin—an effective oral antidiabetes drug—twice daily.

All the patients enrolled in the DPP were clearly at risk for the disease. They had a body mass index of 24 or higher,[1] fasting plasma glucose of 96–125 mg/dL, which is considered impaired fasting glucose, and their glucose tolerance test 2 h after drinking 75 g of a glucose solution was 140–199 mg/dL, which is considered impaired glucose tolerance. With readings like this, it is likely that many would progress to full-blown diabetes mellitus.

The results of the DPP study were as follows: After nearly 3 years, the estimated cumulative incidence of diabetes was 14.4% among patients who adhered to the intensive lifestyle program, 21.6% among the patients who were given metformin, and 28.9% among controls. Put another way, lifestyle modification reduced the incidence of diabetes by 58%, and metformin reduced it by 31% [10]. The problem with these results is that it was not possible to separate responders from nonresponders; in other words, the

---

[1] Body mass index is a person's weight measured in kilograms divided by a person's height in meters squared. Although there are exceptions, readings of 25 or above are considered overweight, readings of 30 or above indicate obesity.

data could not readily be personalized. Sussman and his colleagues designed a diabetes risk model to help resolve this issue.

They started their analysis with several potential risk factors for diabetes that previous research has shown can predict the disease, including the following:

- Fasting blood glucose
- Hemoglobin A1c
- Age
- Body mass index
- Waist-to-hip ratio
- Waist circumference
- Height, sex, race
- Family history of elevated blood glucose
- Smoking status
- Triglyceride levels
- High-density lipoprotein cholesterol
- Systolic blood pressure

To determine which patients obtained the most benefit from metformin and lifestyle modification and which did not, they divided the participants into quarters based on their preintervention risk, using the above list of risk factors to predict their outcomes. They found that seven of the predictive risk factors had the most impact on the risk: fasting blood glucose, hemoglobin A1c, family history of elevated blood glucose, blood triglyceride level, waist measurement in centimeters, and waist-to-hip ratio.

But more importantly, the researchers found "that average reported benefit for metformin was distributed very unevenly across the study population, with the quarter of patients at the highest risk for developing diabetes receiving a dramatic benefit (21.5% absolute reduction in diabetes over three years of treatment) but the remainder of the study population receiving modest or no benefit." By way of contrast, the difference in benefit from the intensive lifestyle training between higher- and lower-risk patients was minimal.

The diabetes risk prediction tool developed by Sussman et al. may help clinicians personalize the care of at-risk patients by identifying those most likely to benefit from drug therapy and spare those who would not from experiencing the adverse effects of metformin, which can include abdominal or stomach discomfort, cough or hoarseness, decreased appetite, diarrhea, fast or shallow breathing, and fever or chills. The prediction tool is available from the University of Michigan's Institute for Healthcare Policy and Innovation and in Fig. 1.2 [11].

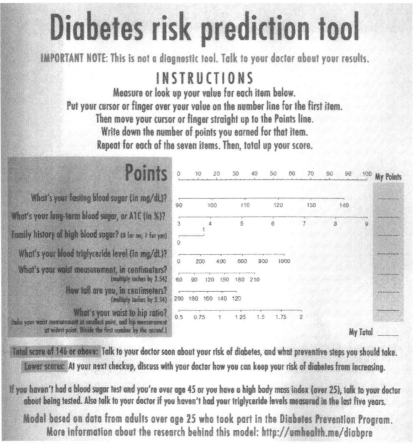

Diabetes risk prediction tool

IMPORTANT NOTE: This is not a diagnostic tool. Talk to your doctor about your results.

INSTRUCTIONS
Measure or look up your value for each item below.
Put your cursor or finger over your value on the number line for the first item.
Then move your cursor or finger straight up to the Points line.
Write down the number of points you earned for that item.
Repeat for each of the seven items. Then, total up your score.

Points   0   10   20   30   40   50   60   70   80   90   100   My Points

What's your fasting blood sugar (in mg/dL)?   90   100   110   120   130   140

What's your long-term blood sugar, or A1C (in %)?   3   4   5   6   7   8   9
                                                              1
                                                         0

Family history of high blood sugar? (0 for no, 1 for yes)

What's your blood triglyceride level (in mg/dL)?   0   200   400   600   800   1000

What's your waist measurement, in centimeters?
(multiply inches by 2.54)   60   90   120   150   180   210

How tall are you, in centimeters?
(multiply inches by 2.54)   200   180   160   140   120

What's your waist to hip ratio?   0.5   0.75   1   1.25   1.5   1.75   2
(take your waist measurement at smallest point, and hip measurement
at widest point. Divide the first number by the second.)   My Total

Total score of 146 or above: Talk to your doctor soon about your risk of diabetes, and what preventive steps you should take.

Lower scores: At your next checkup, discuss with your doctor how you can keep your risk of diabetes from increasing.

If you haven't had a blood sugar test and you're over age 45 or you have a high body mass index (over 25), talk to your doctor about being tested. Also talk to your doctor if you haven't had your triglyceride levels measured in the last five years.

Model based on data from adults over age 25 who took part in the Diabetes Prevention Program.
More information about the research behind this model: http://umhealth.me/diabpre

Figure 1.2 The diabetes risk prediction tool developed by Jeremy Sussman et al. may help clinicians personalize the care of at-risk patients by identifying those most likely to benefit from drug therapy. The researchers point out, however, that the tool has not been verified in clinical practice. *(The chart has been reprinted with the permission of the University of Michigan.)*

In their *New England Journal of Medicine* report documenting the results of the DPP, William Knowler and associates state the following: "An estimated 10 million persons in the United States resemble the participants in the Diabetes Prevention Program in terms of age, body-mass index, and glucose concentrations, according to data from the third National Health and Nutrition Examination Survey. If the study's interventions were implemented among these people, there would be a substantial reduction in the incidence of diabetes." The data from Sussman et al. strongly suggest that this population approach to diabetes prevention would subject millions of patients to unnecessary and potentially harmful drug therapy, but applying a predictive analytics model that includes the 7 risk factors listed above would allow clinicians to target those patients most likely to benefit.

The management of cardiovascular disease is also entering the precision medicine era. The seeds of this precision medicine model were sown in 1948. The Framingham Heart Study, which began to collect data on heart disease risk factors in 1948, has followed the health status of over 5000 adults by performing physical examinations, lab testing, and lifestyle interviews. In 1971, it enrolled another 5000 plus participants—the adult children of the original study subjects, and in 2002, a third generation was added to the cohort. This landmark investigation has identified several contributing risk factors for the disease, including hypertension, hypercholesterolemia, obesity, diabetes, physical inactivity, and smoking [12]. These research data may sound unimpressive in 2017 given the fact that most Americans recognize the health risks associated with a sedentary lifestyle and tobacco, but detecting a relationship between these root causes and heart disease was nothing short of transformational in its day. It is easy to forget that there was a time in the not-too-distant past when anyone opening a popular magazine would read ads that said "More Doctors Smoke Camels Than Any Other Cigarette" or which depicted Santa Claus promoting a popular brand of cigarettes.

Applying the Framingham findings in clinical practice has resulted in dramatic reductions in death and disease burden for cardiovascular diseases in recent decades. But this holistic approach to heart disease now needs to step into the 21st century by adding genetic and genomic data to its collection of risk factors. Recent research on hypertrophic cardiomyopathy (HCM) is a good example of this new direction.

During HCM, the myocardium (heart muscle) becomes hypertrophied, i.e., it thickens. That in turn impairs normal functioning of the heart, making it difficult to pump blood. Signs and symptoms include shortness of breath, chest pain, fainting, and sudden cardiac death. Although the disease can have many causes, familial HCM is inherited as an autosomal dominant trait. One of the defective genes encodes for sarcomere proteins, which are part of the muscle tissue; a mutation in the gene that encodes for the sarcomere proteins produces defective heart muscle.

The Metabolic and Molecular Bases of Inherited Disease explains the following:

Familial or sporadic hypertrophic cardiomyopathy is caused by mutations in one of eight different sarcomere protein genes: cardiac actin; β-cardiac myosin heavy chain; α-tropomyosin; cardiac troponin T; cardiac troponin I; essential myosin light chain; regulatory myosin light chain; and cardiac myosin-binding protein C. At least one other disease gene (mapped to chromosome 7) remains unknown [13].

In an effort to customize treatment of symptomatic obstructive HCM, MyoKardia, Inc, has developed a unique compound designed to target one of the underlying causes, namely the abnormal sarcomere contractibility. The new compound, MYK-461, inhibits this process and has been shown to suppress the disease in animal studies [14]. The Food and Drug Administration (FDA) has granted MYK-461 status as an orphan drug. Of course, promising animal data is no guarantee that the drug will have clinical benefits to patients. But it is one more step in the direction of precision medicine.

UNDERSTANDING THE INTERLOCKING CONTRIBUTORS TO DISEASE

With so many interlocking risk factors and contributing causes, heart disease is one of the most complex disorders to manage. And determining which triggers are contributing to each individual's condition requires an understanding of the *degree* to which each factor is playing a role in each individual. Let's consider a hypothetic patient like Mary Smith as an example. A genetic mutation may be responsible for 70% of her pathology and symptoms, while a diet rich in saturated fat is contributing 5%, a sedentary lifestyle 5%, and a severe chronic stress 20%. John Patrick, on the other hand, may have the same set of signs and symptoms but a different combination of contributing causes that are also different in terms of the degree to which each factor is generating pathology and symptoms. For the sake of this example, let's assume 45% of his cardiac signs and symptoms result from his inability to cope with physical and psychosocial stress, 35% may result from poor diet, and only 20% from a combination of single nucleotide polymorphisms (SNPs) that put him at a disadvantage genetically. Clearly these two patients require different preventive and therapeutic approaches.

Neither profile is completely hypothetical. Familial hypercholesterolemia (FH) is a genetic autosomal dominant disorder that affects 1 in 200–300 persons in its milder heterozygous form [15]. The result is severely elevated plasma levels of low-density lipoprotein (LDL) cholesterol levels—typically 300–400 mg/dL—and premature heart disease [16]. If Ms Smith has FH, the most important contributing cause of her heart disease is a genetic mutation that disrupts the production of the LDL receptor on the surface of her cells. In situations like this, a low-saturated fat diet will only have minimal benefits. An effective treatment strategy would need to target the most important root cause.

A new family of drugs called proprotein convertase subtilisin/kexin type 9 (PCSK9) inhibitors has been able to address this issue in select patients. PCSK9 is a protein involved in the regulation of plasma LDL levels. It appears to control the number of LDL receptors on cell surfaces. The more receptors that appear on the surface of liver cells, the faster the liver can remove LDL cholesterol for the blood stream and the lower the risk of heart disease-promoting hypercholesterolemia. There are currently three new drugs that inhibit PCSK9: evolocumab, alirocumab, and bococizumab. All three have been shown to benefit select patients with FH precisely because they have a mutation in the gene responsible for the production of PCSK9. In these patients, the mutated gene increases expression of PCSK9, which in turn elevates LDL cholesterol levels. For these FH patients, PCSK9 inhibitors are a precise form of medicine.

On the other hand, if Mr. Patrick does not have a genetic disorder, such as FH, but has several environmental causes, his treatment requires a much different approach. Physical and psychosocial stress can contribute to heart disease, as evidenced by several studies [17–19]. By one estimate, psychosocial stress is responsible for 30% of the population attributable risk of acute MI [19]. But individual reactions to stress are highly variable, as reflected by the observation that the amount of the stress hormone cortisol a person produces in response to a triggering event can vary widely. Approximately 17% of individuals do not have diurnal cycles of cortisol as would be expected for the average person, suggesting that the degree to which individuals cope with stress differs from person to person [20]. If Mr. Patrick is a "stress responder," then a stress management program, cognitive behavior therapy, and other methods will likely reduce his risk of clinical signs and symptoms of heart disease if he has not yet developed the disease, or mitigate its signs and symptoms if he has.

An individualized approach may also be required if Mr. Patrick's heart disease is precipitated in part from underlying hypertension. Although a low sodium diet is routinely prescribed to lower a person's blood pressure, sodium sensitivity is not a universal phenomenon. Approximately 25% of Americans are sodium sensitive [21], and Felder et al. report that only 11.8% of hypertensive patients are salt-sensitive [22]. With these statistics in mind, a personalized approach to heart disease might include a diagnostic work-up that establishes the presence of sodium sensitivity. If Mr. Patrick proves to be salt-sensitive, that would justify a salt-restricted diet. If his blood pressure does not increase with a high salt diet, there is no reason to pursue that course.

Unfortunately, the problem with both patient scenarios is that most physicians and nurses in routine clinical practice do not have the tools needed to make many of the decisions required to personalize treatment. The diagnosis of FH is usually missed in primary care practice; measuring blood cortisol levels is rarely performed; there is no readily available test to detect sodium sensitivity, and genetic testing to detect PCSK9 mutations are not commonly available or reimbursable by third party medical insurers. These and other limitations of the precise medicine model will be discussed in Chapter 7.

THE BIOLOGICAL BASIS FOR PERSONALIZED MEDICINE

Although the practical application of the personalized medicine model has yet to be fully realized in many specialties, the biological basis for the model is sound. At the most fundamental level, the population medicine model is functional but flawed. Although all humans have basic biochemical, metabolic, and genetic commonalities—which allow clinicians to treat most patients as though they were an average patient—the variations among individuals are significant and are rarely taken into account when planning prevention and therapy.

One of the reasons a one-size-fits-all approach to patient care is inadequate is because we are not all one size. Opening up a human anatomy textbook gives the impression that all stomachs, livers, and hearts are the same size and shape, even in healthy persons of equivalent weight and height. But opening up actual human bodies indicates otherwise. Similarly, a clinical chemistry reference chart gives the impression that all healthy adults have a serum calcium level between 8.6 and 10.0 mg/dL, for example. But that range, by definition, is only a statistical convention that includes 97.5% of the population tested. The other 2.5% of healthy persons will have readings outside that range—so called false-positive readings. In a US population of 319 million, that means nearly 8 million Americans have blood chemistry levels outside the normal range but remain healthy. In addition, clinical chemistry values are based on a numerical sample of healthy persons, but if that sample is too small, it may not represent the range of the entire US population. By way of example, if serum calcium levels are between 8.6 and 10.0 mg/dL among 1000 healthy adults, what would the range be among 800,000 sampled adults? We do not really know, yet clinicians continue to base diagnostic and therapeutic decisions on the assumption that relatively small sample sizes are representative of the entire population.

The "normal" blood levels of several nutrients have been questions for the same reason. Two early studies that were used to set the Recommended Dietary Allowance for vitamin A in the past, for instance, involved 16 and 8 volunteers, respectively, hardly a representative sample of the human population. Small sample sizes and the biological plausibility of biochemical individuality led Elsas and McCormick to conclude the following: "There are no generalized standards by which minimum or recommended daily allowances (RDA) for essential nutrients can be established for everyone. Rather, there is a continuum within a population of genetically determined variations in requirements extending over a wide range" [23].

Genetic individuality is at the heart of much of the biochemical, anatomical, physiological, and metabolic individuality observed among humans. Exposure to various environmental factors and their interaction with one's genes is likely responsible for the rest. The most extreme examples of biochemical and metabolic individuality brought on by genetic variants have surfaced as inborn errors of metabolism. Phenylketonuria (PKU), for instance, results from a genetic mutation that codes for the enzyme phenylalanine hydroxylase. Without this enzyme, which is needed to convert the amino acid phenylalanine into tyrosine, a person develops PKU, which in turn causes severe mental retardation. But 100 years ago, Archibald E. Garrod, considered by some as the father of precision medicine, pointed out that inborn errors of metabolism are "merely extreme examples of variations of chemical behavior which are probably everywhere present in minor degrees and that this chemical individuality [confers] predisposition to and immunities from the various mishaps which are spoken of as diseases" [24]. Put another way, inborn errors of metabolism exist on a continuum, and the rest of humankind are somewhere on that continuum.

Beebe and Kennedy's review of metabolic individuality states that any individual human genome probably deviates from a reference genome by about 3 million variants [25]. Many of those variants are the results of SNPs (Each of these nucleotide variants involves a change in a single nucleotide. In the deoxyribonucleic acid (DNA) molecule, there are 4 nucleotides: adenine, guanine, cytosine, and thymine, which are repeated over and over in each gene.) (Box 1.1). Many of these genetic variants have been associated with an increased risk of cardiovascular disease, diabetes, kidney disease, hyperlipidemia, and hypertension [26]. More specifically, 13 SNPs have been used as part of a risk score; the score was able to identify individuals that had approximately a 70% higher risk of having an initial coronary heart disease event.

Box 1.1 The ABCs of DNA

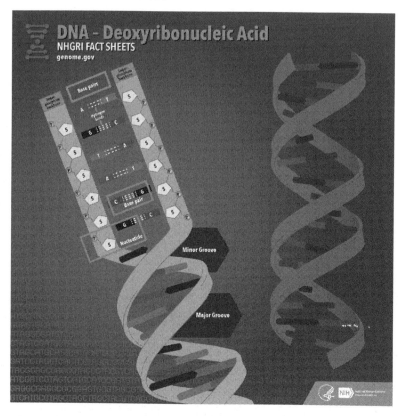

Inherited traits and genetically induced risk factors are carried on chromosomes located in the nucleus of each of our cells. Humans have 23 pairs of these chromosomes, with one of each pair passed on by the mother and one by the father. There are about 30,000 genes located on these chromosomes according to one estimate.

Each gene is a strand of DNA, a helical molecule composed of the sugar deoxyribose, a phosphate group (which consists of phosphorus and oxygen), and nucleotides, which are also referred to as nitrogen bases. The four bases in DNA are adenine (A), thymine (T), guanine (G), and cytosine (C). The four bases are paired off in a regular pattern, with A and T always bonded together and G and C bonded together.

Each strand of DNA contains numerous pairings of A/T and C/G, and it is these bases and their sequence in the DNA molecule that act as a code that determines which proteins the cell will synthesize. During the process of

Continued

Box 1.1 The ABCs of DNA—cont'd

transcription, a DNA strand becomes unzipped, that is, it divides down the middle and is copied into RNA (ribonucleic acid).

RNA is decoded by ribosomes in the cell's cytoplasm, the space outside the nucleus, and translated into amino acids, which in turn are linked together to form a variety of proteins. The body then uses these proteins to carry a wide variety of cell functions. The human body contains 20 amino acids, which are linked in various combinations to form the different body proteins. Synthesis of each amino acid is determined by specific sequences of nucleotides. For example, the amino acid methionine is created when the nucleotides A, T, and G line up during the translation process.

Unfortunately, there are a variety of errors that can occur during this intricate process. One type of mutation, called an SNP, can occur when one of the bases mentioned above is replaced with another. When this occurs, the 3 letter genetic code required to produce specific amino acids is changed, which often results in the required amino acid no longer be manufactured. For instance, if the G (guanine) in the ATG code that translates into the amino acid methionine is replaced with a C (cytosine), the amino acid isoleucine would be made instead. One of 3 letter codes that produces isoleucine is ATC.

Replacing one amino acid with another amino acid can seriously disrupt the functioning of a protein, or disable it altogether. In sickle cell anemia, for instance, GAG mutates to GTG, which in turn causes a replacement of the amino acid glutamic acid with valine. That single substitution changes the structure and function of the hemoglobin molecule, diminishing its ability to carry oxygen to the tissues.

Although many genetic diseases have been traced back to mutations in a single gene, many of the more perplexing disorders that we must cope with have not been linked to a specific gene. Referred to as polygenic, they are likely the result of numerous genetic mutations, combined with environmental factors. Cardiovascular disease, diabetes, and several cancers are believed to fall into this category.

To illustrate how this genetic/biochemical/metabolic continuum affects individuals, consider some examples:

- SNPs have been identified that disrupt the normal utilization of the B vitamin folic acid. Women who carry this mutation are more likely to miscarry if they carry two copies of the defective genetic variant. (Genetic variants are also called alleles) [27].
- The SNP Ala222Va, which affects the gene responsible for the synthesis of the enzyme methylene tetrahydrofolate reductase, not only increases

the risk of neural tube defects (NTDs) and cardiovascular disease but also lowers the likelihood of colon cancer. Individuals with the genetic variant may be at less risk for NTDs and cardiovascular disease if they increase their intake of folic acid. Patrick Stover from Cornell University has suggested that "this may be the best example of a genetic variation that can influence an RDA [Recommended Dietary Allowance] and supports the concept that genetic variation modifies nutrient utilization and potentially dietary requirements."

- Cys282Tyr, an SNP in another gene has been linked to hereditary hemochromatosis, the iron storage disease that causes iron overload and severe liver damage.
- There are polymorphisms that disrupt alcohol metabolism, lactose metabolism, and liver pathways.
- Certain patients taking clopidogrel, the antiplatelet drug, have a genetic mutation that affects a liver enzyme that metabolizes the drug. The enzyme, cytochrome CYP2C19, converts the inactive prodrug into its active form, but when the SNP produces a less effective enzyme, the conversion of the drug is less than optimal, which means it has less anti-platelet effect. Research has demonstrated that patients with this kind of loss-of-function mutation experience more than a threefold increased risk of stent thrombosis and death at 1 year.
- Another genetic variant, HLA-B★15:02, has been associated with a life-threatening reaction to the anticonvulsant drug carbamazepine among Asian patients.
- Patients taking abacavir, which is indicated for HIV infection, are at higher risk of experiencing a hypersensitivity reaction if they carry the HLA-B★5701 gene variant.

Pharmacogenomics and nutritional genomics, the branches of medical science that study these personalized reactions to medications and nutrients, has been a growing area of investigation. Pharmacogenomics has revealed the existence of more than 150 drugs that pose a specific threat to patients with the relevant genetic variants [28]. Among the agents on the FDA list are common drug families such as statins, antibiotics, antidiabetic drugs, heart medications, and anticoagulants. The list includes pravastatin, propranolol, phenytoin, nortriptyline glyburide, paroxetine, sulfamethoxazole and trimethoprim, tamoxifen, telaprevir, tetrabenazine, thioguanine, thioridazine, trastuzumab, tretinoin, trimipramine, valproic acid, venlafaxine, and warfarin.

The FDA points out that pharmacogenomics can play an important role in identifying responders and nonresponders to medications, avoiding adverse events, and optimizing drug dose. Drug labeling may contain information on genomic biomarkers and can describe the following:

- Drug exposure and clinical response variability
- Risk for adverse events
- Genotype-specific dosing
- Mechanisms of drug action
- Polymorphic drug target and disposition genes

One of the more relevant studies that emphasizes the role of genetic individuality in contributing to disease risk and the role of nutrition in exacerbating or ameliorating that risk was performed by Ron Do and associates as part of the InterHeart Study follow-up. The Group analyzed data on over 8000 individuals, comparing the health status of those with several polymorphisms to those without (3820 cases vs. 4294 controls.) More specifically, they studied SNPs located on chromosome 9. (Humans have 23 pairs of chromosomes, a total of 46.) In the p21 section of this chromosome, 4 SNPs have been found, each of which increase the risk of a MI by about 20%. Among individuals who had 2 copies of one of the heart disease-related polymorphisms, there was a twofold increased risk of MI in the InterHeart Study.

They then compared the dietary patterns of cases and controls and discovered that the risk of MI was reduced in those individuals with one of the SNPs if they consumed a diet that was rich in raw fruits and vegetables. In fact, their risk of MI was about the same as in individuals without the genetic mutation when they ate the better diet [29] (Fig. 1.3).

Of course the recommendation to eat more fruits and vegetables is sound advice for everyone and an example of population-based medicine at its best, but many will ignore the advice. Clinicians can add weight to the advice if they can show patients with the genetic markers that they are at greater risk of an MI when they ignore that admonition.

Another genetic marker that may help identify individuals in need of personalized care was described in *Science*. A gene associated with fat mass and obesity has been identified in an analysis of 38,759 study participants during a genome-wide search for type 2 diabetes susceptibility genes. Frayling and associates found that among 16% of adults who had two copies of the gene (referred to as homozygous), their odds of becoming obese was 67% greater than in individuals without the risk allele [30]. The investigators reached their conclusions based on a genome-wide association study that compared 1924 type 2 diabetic patients in Great Britain to 2938 controls; the analysis looked at more than 490,000 SNPs. The obesity-related gene in question,

SNP	Dietary Risk Score	Food Pattern			Intake of Individual Food Items								
		Oriental	Western	Prudent	Meat/ Poultry	Whole Grains	Refined Grains	Deep-Fried Foods	Salty Foods	Fruits	Green Leafy Vegetables	Raw Vegetables	Cooked Vegetables
rs10757274	0.13	0.64	0.19	**0.0017**	0.87	0.60	0.38	0.62	0.056	0.48	0.70	**0.0055**	0.51
rs2283206	0.14	0.86	**0.028**	**0.0004**	0.93	0.53	0.98	0.87	**0.012**	0.28	0.96	**0.0014**	0.21
rs10757278	0.25	0.74	0.11	**0.0022**	0.97	0.75	0.44	0.96	**0.022**	0.47	0.81	**0.0071**	0.25
rs1333049	0.24	0.92	0.17	**0.0011**	0.97	0.36	0.52	0.98	**0.038**	0.36	0.85	**0.0041**	0.48

Interaction tests were performed using logistic regression. Interaction term p-values are shown, after adjustment for main effects of SNP and the dietary variable, as well as ethnicity. The minimum sample size for the analyses is n = 7,989. Associations with p<0.05 are shown in bold. Included individual food items were closely related to the three factors (based on high factor loadings) and were previously shown to be associated with MI. doi:10.1371/journal.pmed.1001106.t003

Figure 1.3 Some single nucleotide polymorphisms (SNPs) apparently work in conjunction with specific food groups to influence the risk of heart disease. (Do R, Xie C, et al. The effect of chromosome 9p21 variants on cardiovascular disease may be modified by dietary intake: evidence from a case/control and a prospective study. PLoS Med 2011;8(10):e1001106.)

labeled FTO, is located on chromosome 16, and the polymorphisms they detected were strongly associated with type 2 diabetes.

THE ROLE OF THE MICROBIOME

When Archibald Garrod conceptualized biochemical individuality, he could not possibly have imagined some of the biological domains now under investigation. Nor could he have envisioned some of the digital tools that are now available to expand clinicians' ability to individualize patient care.

Among the domains that are currently being investigated for their ability to contribute to more personalized care is the microbial content of the human intestinal tract, part of the ecosystem currently referred to as the microbiome. And among the digital tools available to clinicians and the public: smartphone apps, Internet-based patient portals, personal health record programs, wearable-Internet connected sensors.

Scientists have been interested in the role of the intestinal flora in health and disease for centuries. In the 19th century, for example, Eli Metchnikoff of the Pasteur Institute believed that the lactic acid-producing bacteria in the GI tract protected humans from disease and that other microbes produced toxic compounds such as ammonia that contributed to disease. More recently, investigators have begun to study the role of microorganisms on the surface of the body and within the body to determine how they interact with food, drugs, and a variety of metabolites. Humans share several types of bacterial phyla. The human gut contains Firmicutes, Bacteroidetes, Proteobacteria, Actinobacteria, Verrucomicrobia, and Fusobacteria. What distinguishes individuals from one another is the fact that the relative abundance of these bacteria vary widely. Even within an experimental population of healthy young adults free of disease, the diversity is evident [31].

These microorganisms can have a significant impact on one's biochemical makeup. The metabolism of intestinal bacteria produces a by-product called hippuric acid, for example. Concentrations of this compound have been implicated in obesity, diabetes, and inflammatory bowel disease. Illustrating the role of the microbiome in establishing one's biochemical individuality, research on 60 healthy adults found that concentrations of this metabolite vary substantially from person to person and over time. Similarly, urinary levels of hippuric acid seem to vary among ethnic groups. The Japanese, for example, have significantly lower urinary levels of the acid than other groups [31].

Equally important is the observation that drugs can interact with gut microbes, resulting in subtherapeutic and supratherapeutic effects. The lipid-lowering statin simvastatin has been shown to interact with microbial secondary bile acids, resulting in individual variations the drug's ability to lower LDL cholesterol. Animal studies also suggest that suppressing intestinal microbes in the Firmicutes family of bacteria and increasing the dominance of Bacteroidetes can disrupt the pharmacologic effects of the antipsychotic olanzapine. And there is mounting evidence to suggest that the mechanism of action of the popular antidiabetic agent metformin may depend in part on its interaction with gut bacteria.

DIGITAL TOOLS THAT ENABLE PRECISION MEDICINE

Mobile apps, Internet-based patient portals, personal health record programs, and wearable-Internet connected sensors are also capable of personalizing patient care in a way that was not possible in the recent past. And that personalization can take place even in the absence of genetic markers or predictive analytics tools.

In the last few years, Apple has introduced 2 new health-related mobile platforms that are transforming the way researchers, clinicians, and the public interact. ResearchKit is an open-source framework that lets developers create mobile apps, allowing investigators to enroll patients in a variety of studies. The tool enables patients to submit personal information about the health status on their smartphones. These iPhone-enabled apps let users input parameters such as heart rate, mood, exercise data, and much more. Recently, 23andMe designed a tool that enables research subjects to add genetic data to ResearchKit apps as well.

ResearchKit-enhanced studies are providing insights on the environmental triggers that contribute to asthma and helping to refine our understanding of type 2 diabetes by dividing it into several subtypes. The latter is especially relevant in any discussion about personalized medicine because personalized medicine is really about establishing the existence of subpopulations and then targeting therapy to these distinct groups rather than assuming everyone with a disease has the same pathology.

CareKit takes a different approach. Rather than facilitating participants' enrollment in research studies, it provides the public with digital tools to help them and their health-care providers directly manage their conditions. The CareKit platform lets third party developers create apps such as Data Systems OneDrop, for example. The app allows diabetic patients to synchronize blood glucose readings from the glucose meters that are Bluetooth enabled, and from continuous glucose monitoring systems such Dexcom G5, with the Health app on Apple's mobile devices. That data can then be shared with clinicians, who can personalize each patient's regimen as they detect blood glucose patterns and other anomalies. Such anomalies are often not detected by glycated hemoglobin results or other routine lab testing.

Earlier in this chapter, we cited the observations of Robert H. Shmerling, MD, Clinical Chief, Division of Rheumatology at Beth Israel Deaconess Medical Center, who lamented the fact that finding the best medication for a patient with RA is a trial and error process. That is also the case for many other disorders, including clinical depression. A recently introduced mobile app called Iodine Start, which runs on the CareKit platform, was designed to help make the choice of antidepressants less of a trial and error process.

Iodine Start explains to patients what they can expect from their antidepressant, including the amount of time it usually takes for the drug to have a therapeutic effect, and its potential adverse effects. It also provides users with progress reports and helps them monitor their mood and goals in conjunction with their physician. And finally, it helps patients make an informed decision about whether or not the medication is effective and whether it is time to consider another approach. Equally important, the digital tool lets patients compare various antidepressants to one another, comparing their indications, mechanism of action, contraindications, and side effects, including who is more likely to develop serious adverse effects.

For example, in the case of sertraline (Zoloft), the risk is greater for anyone taking anticoagulants or who stops the medication suddenly. Using the detailed drug profiles, along with reviews of other patients who use the Iodine Start app, clinicians and patients may be able to switch to a more appropriate antidepressant in a shorter amount of time—or be reassured that their current drug is the right choice. Mobile apps such as Iodine Start are certainly no substitute for a genetic biomarker or laboratory test that would precisely determine which medication is indicated for each patient. But they can improve the drug selection process and reduce the burden of trial and error for motivated patients.

Motivation, however, is the key to seeing the benefits of Iodine Start and other patient-friendly mobile apps. They require considerable input from users as they track their day-to-day response to medication, their mood, and many other measurable parameters. Such user involvement has advantages and disadvantages. For patients who want more control over their medical care, it is an advantage, but for the many patients who expect their health-care provider to do all the work, these tools are useless. And even for motivated patients, they require an attention to detail and accuracy that many are not accustomed to. Nonetheless, if a patient's health problems are serious enough and they have the cognitive skills and personality to use medical apps appropriately, they can go a long way toward individualizing care. We will discuss the role of mobile apps in precision medicine in more detail in Chapter 5.

CoaguChek Link is another digital tool that can have benefits for users willing to engage. The program is available from Roche Diagnostics to help clinicians caring for patients who are on the anticoagulant warfarin to manage their regimen. The web-based service works in conjunction with a CoaguChek kit that allows patients to do prothrombin time/International Normalized Ratio (PT/INR) testing at home with a single drop of blood. The PT/INR meter is used to report these readings through Roche's Patient Services division. Once again, if the patient is motivated, he or she can help individualize their care by reviewing the test result history, trends, and to determine how often their readings remain in the therapeutic range, by taking advantage of a Roche website.

REFERENCES

[1] Kirkham B. Tumor necrosis factor-alpha inhibitors: an overview of adverse effects. UpToDate; April 26, 2016. https://www.uptodate.com/contents/tumor-necrosis-factor-alpha-inhibitors-an-overview-of-adverse-effects.
[2] Schork NJ. Personalized medicine: time for one-person trials. Nature April 30, 2015;530:609–11. http://www.nature.com/news/personalized-medicine-time-for-one-person-trials-1.17411#/imprecision.
[3] Huffman TF, et al. Statins for the primary prevention of cardiovascular disease. Cochrane Library; January 31, 2013. http://www.cochrane.org/CD004816/VASC_statins-for-the-primary-prevention-of-cardiovascular-disease.
[4] CAPRIE Steering Committee. A randomised, blinded, trial of clopidogrel versus aspirin in patients at risk of ischaemic events (CAPRIE). Lancet 1996;348:1329–39.
[5] Raslan HM, et al. Association of PTPN22 1858C→T polymorphism, HLA-DRB1 shared epitope and autoantibodies with rheumatoid arthritis. Rheumatol Int August 2016;36:1167–75.
[6] National Library of Medicine. Genetics home references: rheumatoid arthritis, genetic changes. June 21, 2016. https://ghr.nlm.nih.gov/condition/rheumatoid-arthritis#genes.
[7] Gan RW, et al. Omega-3 fatty acids are associated with a lower prevalence of autoantibodies in shared epitope-positive subjects at risk for rheumatoid arthritis. Ann Rheum Dis 2016. http://dx.doi.org/10.1136/annrheumdis-2016-209154. Published online May 17, 2016.
[8] Precision Medicine Initiative (PMI) Working Group. The precision medicine initiative cohort program – building a research foundation for 21st century medicine. September 17, 2015. http://acd.od.nih.gov/reports/DRAFT-PMI-WG-Report-9-11-2015-508.pdf.

[9] Sussman JB, Kent DM, Nelson JP, et al. Improving diabetes prevention with benefit based tailored treatment: risk based reanalysis of diabetes prevention program. BMJ 2015;350:h454. http://www.bmj.com/content/350/bmj.h454.

[10] Knowler WC, Barrett-Connor E, Fowler SE, Hamman RF, Lachin JM, Walker EA, et al. Reduction in the incidence of type 2 diabetes with lifestyle intervention or metformin. N Engl J Med 2002;346:393–403.

[11] Institute for Healthcare Policy and Innovation, University of Michigan. Precision medicine to prevent diabetes? Researchers develop personalized way to steer prevention efforts. February 19, 2015. http://ihpi.umich.edu/news/precision-medicine-prevent-diabetes-researchers-develop-personalized-way-steer-prevention-efforts.

[12] Framingham Heart Study. History of the Framingham heart study. https://www.framinghamheartstudy.org/about-fhs/history.php.

[13] Seidman CE, Seidman JG. Hypertrophic cardiomyopathy. In: Valle D, editor in Chief. The metabolic and molecular bases of inherited disease. Online ed. New York: McGraw-Hill; 2007.

[14] Green EM, Wakimoto H, et al. A small-molecule inhibitor of sarcomere contractility suppresses hypertrophic cardiomyopathy in mice. Science 2016;351:617–21. http://science.sciencemag.org/content/351/6273/617.

[15] Goldberg AC, Gidding SS. Knowing the prevalence of familial hypercholesterolemia matters. Circulation March 15, 2016. http://dx.doi.org/10.1161/CIRCULATIONAHA.116.021673http://circ.ahajournals.org/content/133/11/1054.

[16] Marian AJ, Brugada R, Roberts R. Cardiovascular diseases caused by genetic abnormalities. In: Hurst's the heart. 13th ed. New York: McGraw-Hill; 2011.

[17] Cohen BE, Panguluri P, et al. Psychological risk factors and the metabolic syndrome in patients with coronary heart disease: findings from the heart and soul study. Psychiatry Res 2010;175:133–7. http://www.ncbi.nlm.nih.gov/pubmed/19969373.

[18] Ohira T. Psychological distress and cardiovascular disease: the circulatory risk in communities study (CIRCS). J Epidemiol 2010;20(3):185–91. http://www.ncbi.nlm.nih.gov/pubmed/20431233.

[19] Roemmich JN, Fedda DM, et al. Stress-induced cardiovascular reactivity and atherogenesis in adolescents. Atherosclerosis 2011;215(2):465–70. http://www.ncbi.nlm.nih.gov/pubmed/21296350.

[20] Smyth JM, Ockenfels MC, et al. Individual differences in the diurnal cycle of cortisol. Psychoneuroendocrinology 1997;22(2):89–105.

[21] Gildea JJ, Lahiff DT, et al. A linear relationship between the ex-vivo sodium mediated expression of two sodium regulatory pathways as a surrogate marker of salt sensitivity of blood pressure in exfoliated human renal proximal tubule cells: the virtual renal biopsy. Clin Chim Acta 2013;5:421–42.

[22] Felder RA, White MJ, et al. Diagnostic tools for hypertension and salt sensitivity testing. Curr Opin Nephrol Hypertens 2013;22(1):65–76.

[23] Cerrato P. Recommended dietary allowances for vitamins. JAMA 1988;260:1242.

[24] Suhre K, Raffler J, et al. Biochemical insights from population studies with genetics and metabolomics. Arch Biochem Biophys 2016;589:168–76.

[25] Beebe K, Kennedy AD. Sharpening precision medicine by a thorough interrogation of metabolic individuality. Comput Struct Biotechnol J 2016;14:97–105.

[26] Gottesman O, Bottinger E. Personalized medicine in clinical practice. In: Murray MF, Babyasky MW, et al., editors. Clinical genomics: practical applications in adult patient care. New York: McGraw-Hill Education; 2014.

[27] Stover PJ. Influence of human genetic variation on nutritional requirements. Am J Clin Nutr 2006;83(Suppl.):436S–42S. http://ajcn.nutrition.org/content/83/2/436S.full.

[28] Food and Drug Administration. Table of pharmacogenomic biomarkers in drug label-ing. July 11, 2016. http://www.fda.gov/Drugs/ScienceResearch/ResearchAreas/Pharmacogenetics/ucm083378.htm.

[29] Do R, Xie C, et al. The effect of chromosome 9p21 variants on cardiovascular dis-ease may be modified by dietary intake: evidence from a case/control and a prospec-tive study. PLoS Med 2011;8(10):e1001106. http://journals.plos.org/plosmedicine/article?id=10.1371/journal.pmed.1001106.

[30] Frayling TM, Timpson NJ, et al. A common variant in the FTO gene is associ-ated with body mass index and predisposes to childhood and adult obesity. Science 2007;316:889–94.

[31] Patterson AD, Turnbaugh PJ. Microbial determinants of biochemical individuality and their impact on toxicology and pharmacology. Cell Metab November 4, 2014;20:761–8.

CHAPTER TWO

Precision Medicine Initiatives and Programs

As we mentioned in Chapter 1, the US government has launched a major initiative to develop precision medicine into a workable model that can eventually be applied in day-to-day clinical practice. The $215 million commitment is only one of many initiatives and programs now being developed throughout the world.

The United Kingdom, for example, has a program called Genomics England, a genome sequencing project that will analyze the genes of 100,000 individuals with rare inherited diseases, cancer, and infection [1]. Similarly, the German Federal Ministry of Education and Research has introduced an Action Plan called "Individualized Medicine: a New Way in Research and Healthcare." Since then, the country has invested more than 40 million euros in personalized medicine [2]. China is also developing a program in this area. The China Precision Medicine Initiative (PMI) is a 15-year project that will invest the equivalent $9.1 billion US dollars [3].

In addition to government-related projects, there are numerous university-based programs that are now focusing on precision medicine. Among the medical school-affiliated programs that concentrate on precision medicine: University of California at San Francisco, the Mayo Clinic Center for Individualized Medicine, Columbia University's Institute of Genomic Medicine, Baylor Precision Medicine Institute, the Department of Biomedical Informatics at Harvard Medical School, the Johns Hopkins Individualized Health Initiative, Precision Medicine at Stanford Medicine X, and the Penn Center for Precision Medicine. The purpose of this chapter is to discuss some of these public and private initiatives to give readers an overview of what thought leaders in the field are doing to actualize the promise of precision medicine.

Realizing the Promise of Precision Medicine
ISBN 978-0-12-811635-7
http://dx.doi.org/10.1016/B978-0-12-811635-7.00002-6

THE US PRECISION MEDICINE INITIATIVE

During President Barack Obama's 2015 State of the Union Address, he announced the PMI, which he said would "bring us closer to curing diseases like cancer and diabetes, and to give all of us access to the personalized information we need to keep ourselves and our families healthier" [4].

The Initiative not only hopes to develop an understanding of health and disease that leads to more accurate diagnosis, more effective preventive programs, and better treatment but it also hopes to change the culture of medical practice, engaging individuals not just as patients or research participants but as *active partners* in their own care.

The PMI is being funded by $215 million, which is being invested in the National Institutes of Health (NIH), the Food and Drug Administration (FDA), and the Office of the National Coordinator for Health Information Technology (ONC). $130 million is being allocated to develop a national research cohort of a million or more volunteers who will serve as research participants. Seventy million dollars will be set aside for the National Cancer Institute to help identify genomic drivers of cancer and to use these findings to improve cancer therapy. FDA will receive $10 million to assist in the development of databases that support the regulatory framework needed to make precision medicine a reality, and ONC will receive $5 million to further develop interoperability standards and to strengthen the private secure exchange of participants' health data [5].

Unlike many initiatives proposed by the Obama Administration, PMI has received support from both major political parties in the United States, with the US Congress's appropriations committees in both houses approving funds as part of a spending bill for 2017 [6].

The Precision Medicine Initiative Cohort Program (PMI-CP)—now called the All of Us Research Program—is by far the most ambitious component of the program (Fig. 2.1). It will require the recruitment of hundreds of thousands of Americans from all walks of life and aims to represent the entire US population. The group will have to agree to share their health data, provide biospecimens, and remain in contact with the investigators managing the program over many years. The report from the PMI working group envisions recruiting these volunteers in two ways: One subgroup will volunteer directly, while the other subgroup will be recruited through health provider organizations (HPOs). Both groups will also agree to have a baseline health examination performed. The current goal is to recruit at least 1 million participants

Figure 2.1 The United States Precision Medicine Initiative Cohort Program is best symbolized by the cover of the 2015 report from the PMI Report, which illustrates the need to personalize care for a diverse population of young and old, male and female, and healthy and infirm.

over approximately 4 years. Volunteers recruited through health providers receive care over time and typically have a longitudinal record of care, which will likely be available in an electronic format with ongoing, documented follow-up. Among the settings from which the PMI project can draw: academic medical centers, Federally Qualified Health Centers, vertically integrated private health-care organizations (e.g., Kaiser Permanente), and vertically integrated governmental organizations (e.g., the Veterans Administration (VA)).

The VA will play an important role in the PMI project according to officials. The NIH plans to collaborate with the VA to make sure former servicemen and women have a chance to participate in the PMI-CP. The VA already has a Million Veteran Program (MVP) in place that is a voluntary research program that partners with veterans already receiving care from VA facilities. Since it is studying how genes affect the health of the population, it seems a perfect match for inclusion in a precision medicine project that places a great deal of emphasis on genetic individuality and its role in health and disease. The goal of the MVP is to collect and store blood samples and health information from 1 million veterans and use it to investigate diseases such as diabetes, cancer, and military-related illnesses, including posttraumatic stress syndrome.

Other participant groups that the PMI project may look to include research cohorts engaged in clinical trials, including the Framingham Heart Study and Multiethnic Study of Atherosclerosis, community groups, and patient and health advocacy groups. But it is more likely that these three groups will play a role in promoting the recruitment of individuals as PMI cohort participants, either as direct volunteers (those with no affiliation to a PMI HPO, described below) or through their involvement with provider organizations. The Working Group emphasizes their value by stating the following: "Leveraging the energy and member relationships of these non-HPOs to encourage their members to become involved in the PMI cohort would form an important part of a recruitment strategy, especially with more diverse communities."

The project managers also envision the participants getting involved in other ways, providing feedback, offering input during the planning and implementation of the project, and even being represented on the PMI-CP governance and oversight committees. This is a rational approach to subjects' participation because it can help ensure trust, which is a key component of a program in which the federal government and affiliated researchers are asking US citizens for a wide array of personal information on their health. Many Americans are suspicious of the government's ability to keep their data secure and private, and there is some justification for this concern considering recent breaches of government databases.

MANAGING DATA

Collecting, storing, and analyzing the massive amounts of health data will require well-proven and innovative technologies. As the PMI report explains the issue:

Efficient structuring and management of data is important to the success of the PMI-CP. Toward this end, the Working Group recommends use of a common data model to organize data similarly across HPOs and from direct volunteers, where possible, while recognizing that many data types useful from clinical investigation may not easily be transformed to an existing or created standard at this time. The best approach will balance normalization of only the highest value data initially for all participants followed by on-demand data curation of other data as driven by scientific demand. In addition, the Working Group recommends that existing data standards and common data models be leveraged where possible, while recognizing that standards do not exist for many emerging modalities, such as a number of sensor technologies. The Working Group recommends early selection of commonly used mHealth technologies to gain experience in use and integration of these new modalities.

To facilitate data access, data normalization, and participant engagement, the Working Group recommends that the PMI-CP follow a 'hub-and-spoke' model that has a Coordinating Center to provide a single point of contact for coordinating data, biospecimens, participant communication and engagement, and research studies. The Working Group encourages NIH to consider novel collaborations with not-for-profit and commercial organizations to achieve state-of-the-art analysis methods, scientific rigor, elastic storage and compute capabilities, and technological expertise. For data storage and access, the Working Group recommends the PMI-CP pursue a hybrid data and analytics architecture that leverages both centralized data storage of core data while preserving federated access to additional data at the nodes across the network, as needed by specific studies. This hybrid model would accelerate execution of many research queries but still allow detailed data access for queries not addressable through the current data common data models.

Collecting and storing the tissue samples of hundreds of thousands of participants poses many challenges. The PMI managers have chosen the Mayo Clinic in Rochester, Minnesota, to build this biobank. These data will be combined with a large variety of lifestyle and medical information that volunteers provide to help investigators determine the individual differences that contribute to health and disease.

EXECUTING THE US PRECISION MEDICINE INITIATIVE

Some observers may question the need for a million person research project, especially when one considers the complex technological and governance issues involved in managing so many volunteers and the massive amounts of data they will generate. But the rationale for this large cohort becomes clear when one looks at the mathematics.

To arrive at a useful cohort size, the PMI Working Group examined electronic health record (EHR) data from a large EHR-associated biobank to estimate disease frequencies in 1 million Americans. It pulled ICD (International Classification of Diseases) codes from individuals participating in the Vanderbilt BioVU DNA biobank and then mapped these data into a disease classification system that included over 1600 disease groupings. These data were also compared to data from the Electronic Medical Records and Genomics Network, which is part of the NIH's National Human Genome Research Institute. The latter contained data from individuals with particular diseases.

Using this approach, the Working Group determined that in a group of 1 million participants, one could expect to detect 14,981 incidents of myocardial infarction at 5 years and 27,112 cases in 10 years. Similarly, one can expect to find more than 4000 and 11,000 cases of epilepsy in these 2 time frames and over 2000 and 4000 cases of Parkinson's disease, respectively.

The Working Group also calculated the statistical power that needs to be reached to establish a firm association between a genetic or environmental risk factor and a specific disease. For example, if one tests an exposure to a risk factor such as secondhand tobacco smoke, which is present in 10% of the volunteers, looking for an association with a disease that occurs in 25,000 cases, applying an alpha of 10^{-7}, one can expect 80% power to detect an odds ratio of at least 1.16.

In the final analysis, a research group composed on 1 million volunteers will be well powered to detect relationships between exposures and hundreds of health outcomes.

Of course, any discovered associations between genetic and environmental risk factors and diseases will have little meaning to the US population as a whole if the 1 million participants have been carelessly selected. To avoid that possibility, the recruiting process will need to recruit Americans from diverse social, racial, ethnic, and ancestral populations, and from various geographies, social environments, and economic situations. Currently, most of the research data on which standard treatment protocols and clinical practice guidelines are based has been extrapolated from a population of white, male, urban, and relatively affluent individuals. This relatively homogenous group is hardly representative of the entire US population and is becoming less and less representative as more immigrant groups make up a larger portion of the population. The 2014 US Census indicates there are more than 20 million children under the age of 5 years living in America; 50.2% are minorities. Hispanics now make up the largest minority [7].

A national sample of African Americans surveyed by Chanita Hughes Halbert and colleagues suggests that recruiting a truly representative cohort for the PMI project will certainly be challenging. Using telephone-based survey, Halbert et al. specifically asked respondents about how likely they would be to participate in a study that was sponsored by the government, that would gather information about the health of African Americans, would require answering questionnaires, and would require researchers to do a cheek swab. The survey also pointed out that the data would also be used in future studies, and that the results would not be shared with participants. Only one third of the respondents said they would participate [8]. When asked specifically about whether they would take part in the experiment "If I did not know who would be able to obtain my personal information (e.g., blood or saliva sample, medical history)," 60% said they would be unlikely to participate. (Additional reasons that survey respondents had for not participating in PMI are spelled out in Table 2.1.)

Table 2.1 Reasons for participating in cancer genetics research

Participation reason	Unlikely (%)
If I did not know who would be able to obtain my personal information (e.g., blood or saliva sample, medical history).	60
If it were hard for me to get to where the study was being conducted.	63
If findings from the study would not be made available to me.	59
If I had to participate in the study for a long period of time.	48
If the results of the study would be used to develop drugs I might not be able to afford.	40
If I did not receive any financial compensation for my time.	38
If my family members or friends told me not to participate in the study.	36
If my information (e.g., medical history, blood or saliva sample) would be used in other studies.	34
If I were afraid of getting information about my health that I did not want to know.	32
If the study were sponsored by a pharmaceutical company.	30
If the study only included African Americans.	26
If I had to participate in the study for a short period of time.	13
If my family members or friends did not think I should participate in the study.	13
If the study addressed a health condition I was worried about.	13
If someone from my racial group was conducting the study.	12
If I would get free medication or health care.	11

From Halbert CH, McDonald J, Vadaparampil S, Rice L, Jefferson M. Conducting precision medicine research with African Americans. PLoS One 2016;11(7):e0154850. http://dx.doi.org/10.1371/journal.pone.0154850. http://journals.plos.org/plosone/article?id=10.1371%2Fjournal.pone.0154850.

The fact that the hypothetical scenario the researchers presented to this group of African Americans was very similar to the actual parameters being used in the actual PMI project is discouraging.

 ## WHAT WILL THE PRECISION MEDICINE INITIATIVE REVEAL?

The PMI Working Group outlined eight major endpoints that it hopes to reach:

1. Development of quantitative estimates of risk for a range of diseases by integrating environmental exposures, genetic factors, and gene–environment interactions.
2. Identification of the determinants of safety and efficacy for commonly used therapeutics.

3. Discovery of biomarkers that identify individuals with an increased risk of developing common diseases.

4. The use of home and mobile health (mHealth) technologies to correlate body measurements and environmental exposures with health outcomes.

5. Determination of the clinical impact of loss-of-function mutations.

6. Development of new disease classifications and relationships.

7. Empowerment of participants with data to improve their own health.

8. Enrollment of PMI cohort participants in clinical trials of targeted therapies.

The Working Group has also created an ambitious timeline for achieving these accomplishments, which we have reproduced as Table 2.2.

Each of these endpoints would have a profound impact on patient care if they can be reached. Decades of epidemiological studies and randomized clinical trials have established numerous risk factors and contributing causes for a variety of disorders, but we have only scratched the surface. And in many cases, the association between risks, causes, and diseases are tenuous because sample sizes have been too small or because 95% confidence intervals have been borderline. The PMI Report explains the following: "The PMI cohort will provide a broadly useful resource for rigorously validating and quantifying the contributions of genetic and environmental risk factors,

Table 2.2 Timeline when expected Precision Medicine Initiative cohort capabilities will be realized

Cohort capabilities	Time in years			
	0–2	3–5	5–10	>10
1. Discovery of disease risk factors	+	+++	+++	++++
2. Pharmacogenomics	+	+++	+++	+++
3. Discovery of disease biomarkers	+	++	+++	+++
4. mHealth connections with disease outcomes		+	++	++++
5. Impact of loss-of-function mutations		+	+++	+++
6. New classifications of diseases		+	+++	++++
7. Empowering participants	+++	+++	+++	+++
8. Clinical trials of targeted therapies		+	+++	+++

The estimated timeline for focused research for each type of investigation is indicated by the number of "+" characters in each cell. *mHealth*, mobile health.
From The precision medicine initiative cohort program – building a research foundation for 21st century medicine: precision medicine initiative (PMI) working group report to the advisory committee to the director, NIH; September 17, 2015. http://acd.od.nih.gov/reports/DRAFT-PMI-WG-Report-9-11-2015-508.pdf.

as well as their interactions with one another, in a large, diverse population. This will certainly include risk factors that have been proposed from smaller studies, but the comprehensiveness of the PMI cohort dataset will also allow for data scientists to identify new and unexpected associations. As it grows in breadth and depth, the PMI cohort will allow for these estimates on uncommon as well as common diseases."

Reaching Endpoint #2, which aims to more precisely determine how to safely and effectively use commonly available medications, is sorely needed. It is estimated that approximately one out of every two American adults takes at least one drug and 22% take three or more. And the sheer number of patients who experience serious adverse effects from prescription medication, over the counter (OTC) drugs, nutritional supplements, and herbal remedies is staggering. As we have pointed out in the last chapter, research conducted to prove the efficacy and safety of drugs in the United States usually establishes their value for the population as a whole. This research does little to identify responders and nonresponders, not does it pinpoint individuals who have a genetic or environmentally induced predisposition to adverse reactions. Approximately 4.5 million office visits occur annually because of adverse reactions, and these reactions result in thousands of hospitalizations. The PMI project hopes to prevent many of these reactions by discovering genomic and environmental factors that trigger them. The ultimate goal says the PMI Working Group report is "By allowing clinicians to tailor medication and dosage to each individual's profile, these discoveries have the potential to increase the efficacy of prescribed therapeutics, to reduce the incidence of adverse effects, to improve health outcome and to reduce overall healthcare costs."

THE AGE OF BIOMARKERS IS UPON US

PMI also has its sights set on understanding as yet undiscovered biomarkers capable of identifying individuals at risk for specific disorders. We now have inexpensive genome and cell-free DNA sequencing tools available that can measure DNA, RNA, a wide range of metabolites, and signaling molecules, and we can also measure immune system activity that may signal the preclinical stage of disease.

Traditional biomarkers vary widely in their value as predictive tools for the detection of disease. Prostate-specific antigen (PSA), for example, has proven unreliable at predicting prostate cancer, prompting the US Preventive Services Task to recommend against its routine use as a screening test for the disease in men 50 years and older [9]. Antibody markers such as anti-citrullinated protein antibodies, on the other hand, are more useful in the management of rheumatoid arthritis. And metabolites such as phenylalanine in urine of infants are invaluable in the diagnosis of phenylketonuria. But recent advances in cancer medicine and genomics suggest that there are several new biomarkers worth exploring [10].

Although biomarkers have been discussed as valuable aids in helping to individualize care, more often than not, these markers help identify *subgroups* rather than single individuals who will benefit from the detection of said marker. For example, the presence of epidermal growth factor receptor (EGFR) has been shown to help distinguish patients with lung adenocarcinoma from one another. The subgroup of patients with EGFR-positive nonsmall cell lung adenocarcinoma can now be distinguished from those with general lung adenocarcinoma, and that in turn makes them candidates for a different chemotherapy regimen than the majority of patients with the malignancy. One of the ways in which these recently discovered biomarkers differ from the aforementioned traditional markers is "the magnitude of data collected and the speed at which data from different sources is (usually simultaneously) analyzed," according to Vargas and Harris [10].

These biomarkers may not only alter the type of treatment patients receive but they can also facilitate more accurate prediction in at risk patients and improve diagnosis and prognosis. Such markers can be correlational in nature or functional in nature. No doubt the Million person PMI cohort will reveal numerous associations between risk factors and disease. And many of the biochemical markers being measured in the program will be correlated with increased incidence of specific diseases. But there will also be occasions in which the biomarker/disease correlation will be functional in nature, i.e., there will be an identified mechanism of action linking the biomarker to the disorder.

Another role for biomarkers that will benefit from the data generated by PMI is sometimes referred to as companion diagnostics. Currently, the US FDA has a dedicated pathway for drug approvals that makes use of biomarkers paired with a companion therapeutic agent. A case in point is the aforementioned EGFR marker, which has been paired with the drug afatinib (Gilotrif), an EGFR inhibitor. The drug has been approved as first-line treatment for metastatic nonsmall cell lung cancer with common EGFR mutations.

The study of biomarkers has become quite sophisticated in recent years, and a rudimentary understanding of this field is valuable for clinicians, technologists, and business executives interested in working in this area. Both single biomarkers and biomarker panels consisting of more than one marker have been developed. These markers can result from measurement of several types of physiological and chemical parameters. The EGFR mutations fall into the genomic category. Other biomarkers are being identified by measuring transcriptomics, proteomics, microbiomes, and other "omics."

Other genomic-derived biomarkers that now have a role to play in clinical medicine include anaplastic lymphoma receptor tyrosine kinase (ALK) translocations. The ALK gene instructs cells to synthesize a protein called anaplastic lymphoma kinase. This enzyme is part of a complex pathway involved in cellular growth, division, and maturation. It is not hard to imagine that a genetic defect involving this pathway could contribute to cancer, which is a form of cell division gone awry.

A gene translocation occurs when sections of DNA jump around to places on a chromosome where they do not belong. ALK translocations can be detected by a diagnostic process called fluorescent in situ hybridization and that in turn identifies patients who would likely benefit from ALK-inhibiting drugs such as crizotinib, which is used to treat advanced nonsmall cell lung cancer.

Proteomics, another source of biomarkers, refers to the study of body proteins made possible through new technologies such as tissue microarrays and advances in spectrophotometry. In the literature review by Vargas and Harris, they point out that 17 circulating proteins have been identified that are biomarkers for nonsmall cell lung cancer. Researchers have also identified circulating inflammatory proteins that have been shown to have clinical value in the prognosis of lung cancer.

Endpoint #4, which aims to make fuller use of mHealth technologies to link body measurements and exposure to various environment triggers with health outcomes, seems a more doable goal to reach in our view—at least in the near term. Although biomarker research holds promise and has yielded a limited number of useful diagnostic and treatment tools, it relies on very specialized measurement technologies—spectrophotometry, for instance. Such tools are not readily available in the clinic. Such research is also very expensive.

In contrast, mHealth technology is more "down to earth" and can generate valuable datasets that can have immediate benefits to patients. Mobile apps can now measure heart rate and rhythm, perform EKGs, measure some aspects of mental health, and track sleep behavior and numerous other parameters. And these data can be collected at very low cost using smartphones that now fill the pockets of millions of Americans. Examples of the value of mHealth technologies illustrate their value: Typing speed has been shown to predict the development and treatment efficacy in Parkinson's disease [11]. And sensor-detected paroxysmal atrial fibrillation may help identify those at increased risk of embolic stroke [12]. In Chapter 5, we will discuss the role of mobile technology in precision medicine in more detail.

PATIENT ENGAGEMENT: A CRITICAL COMPONENT OF PRECISION MEDICINE

As the title of our book suggests, engaging patients and consumers in self-care and in their medical treatment is one of the most important keys to the success of precision medicine. It is also a key component to the PMI project sponsored by the US government.

In 2016, NIH awarded $55 million to several organizations to help build the infrastructure for PMI and to engage the public more fully in the project. Five groups—Scripps Research Institute, Vibrent Health, PatientsLikeMe, Sage Bionetworks, and Walgreens—will make up the Participant Technologies Center, which will be involved in creating the mobile applications needed to enroll and collect volunteer data and to communicate with participants [13].

A recent survey of more than 2600 individuals found that 79% believe the PMI is a good idea, and more than half of the respondents say they would "definitely or probably participate in a study if asked." Support for the study varied little among demographic groups, but respondents with higher levels of education were more likely to take part in the Initiative if they were recruited. When asked what would motivate them the most when considering participation in the study, the strongest incentive cited was "learning information about their health." In fact, Fig. 2.2 shows that enthusiasm extended to a long list of data points, including lab results, information about the genome, and nutrition information.

Of course, it is easy to express support for a research project on paper but quite a different matter to actually participate. Thus there is a need to actively recruit interested parties, which is what the latest NIH grant will help accomplish. The Scripps Research Institute's goal, for example, is to enroll 350,000 of the 1 million volunteers requested by the NIH.

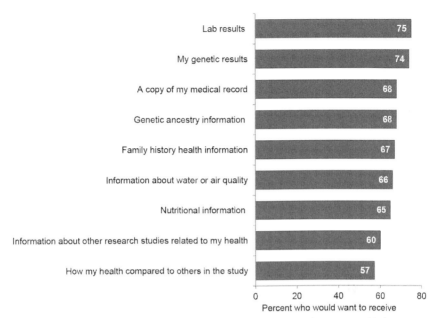

Figure 2.2 If the study went forward, as a study participant what types of information would you like to receive? *(From Kaufman DJ, et al. A survey of U.S. adults' opinions about conduct of a Nationwide Precision Medicine Initiative® cohort study of genes and environment. PLoS One August 17, 2016.* http://journals.plos.org/plosone/article?id=10.1371/journal.pone.0160461.)

As we mentioned earlier in this chapter, the 1 million volunteers who will make up the PMI are to be recruited from health-care provider organizations and directly from the general public. The Participant Technologies Center is the focal point for recruiting the latter group. In addition to this function, the Center "will also develop, test, maintain and upgrade, as needed, PMI Cohort Program mobile applications. These mobile apps will be used to enroll, consent, collect data from and communicate with PMI Cohort Program participants. The center also will develop parallel platforms to deliver these same functions to those without smartphones, and work with various technology organizations to increase smartphone accessibility" [14].

Despite the federal government's interest in recruiting "nonpatients" from the general public, the PMI Working group still maintains that recruiting participants from provider organizations will probably result in the fastest and most cost-efficient route to enrolling large numbers of participants who already have a treasure trove of health-care data that can be analyzed. It also believes this same population is most likely to generate easily accessible follow-up health data. In the words of the PMI Working Group report:

> Health provider organizations (HPOs) with primary access to and the ability to share ongoing, comprehensive health data have exceedingly strong potential to be highly effective partners with the PMI-CP, functioning as sites or 'nodes' within the PMI cohort for recruitment, communication, biospecimen collection, and healthcare data collection (through their clinical care relationship). The Working Group expects that the majority of the participants will be recruited from HPOs. Diversity and inclusion of historically understudied populations should be considerations in selecting HPO cohorts for inclusion in the PMI cohort, as described above. The HPOs may continue to be the primary contact with the volunteers they recruit from within their patient population. Recruitment of the individuals, collection of biospecimens, and regularization and transmittal of participant EHR data represent significant, primary efforts to build a valuable PMI cohort. These organizations will need funding to support these efforts.

THE VALUE OF PRECISION MEDICINE INITIATIVE DATA REACHES BEYOND THE US PRECISION MEDICINE INITIATIVE PROJECT

While one of the primary goals of the PMI project is to generate insights into the causation of complex, polygenic diseases, the data collected during this national experiment can also serve as an invaluable resource for the health-care organizations participating in the project. The PMI managers envision sharing the biologic and phenotype data they collect with participating providers and researchers, realizing that this database could serve as a rich source for local HPOs as they embark on their own clinical and epidemiological research projects.

Of course, before we can entertain the use of PMI data outside of the project itself, PMI managers first have to make certain that the collection, storage, and analysis of its data are optimal. As the table in Chapter 1 of this book illustrates, PMI has set out on a grand mission. In addition to the usual datasets, including demographics, personal consent, baseline health exam data, and structured clinical data from EHRs, the goal is to explore parameters rarely measured in routine ambulatory patient visits or in-hospital stays. This is where PMI shines because gathering these data can reveal as yet unimagined risk factors for a wide array of disorders.

Take as an example the collection of self-reported measures, including quality of life and well-being evaluations, gender identity, or functional capabilities. It is conceivable that certain standard surgical or pharmacologic interventions result in major declines in patient well-being, despite the fact that randomized clinical trials have documented fewer disease-specific signs and symptoms and improved lab readings. On the other hand, it is entirely possible that subgroups of volunteers with certain types of physical or psychological trauma may report improved functional capacity without *any* medical intervention, the result of the simple passage of time and the body's innate ability to heal itself. Such research revelations would give new meaning to the expression: Time heals all wounds. Collecting and analyzing self-reported measures about quality of life may very well reveal these types of insights.

The list of parameters to be charted by PMI also includes a variety of behavioral and lifestyle measures, including diet, physical activity, alternative therapies, smoking, alcohol, and assessment of known risk factors (e.g., guns, illicit drug use). The investigators hope to find correlations between these parameters and clinical events, drug response, and health outcomes.

One of the most interesting collections of measures that have the potential to uncover new risk factors for disease is sensor-based observations gathered from smartphones, wearables, and home-based devices. The number and variety of mHealth apps now available for iPhone and Android devices seems endless. Collecting and analyzing these data will likely turn up a number of surprises. There are now apps and accompanying phone attachments that can monitor blood pressure, oxygen saturation in the blood, respiratory rate, and cardiac rate. Others can perform an EKG, measure body temperature, and evaluate skin conductance—which has been correlated with the stress response.

The PMI team also wants to collect and evaluate a variety of biospecimen-derived laboratory data, including genomics, proteomics, metabolites, cell-free DNA, single cell studies, infectious exposures, standard clinical chemistries, and histopathology. Currently, many of the leading oncology treatment centers are doing genome sequencing of patients and tumors to look for biomarkers that may help individualize treatment, a topic we will discuss in Chapter 3. But that type of analysis only scratches the surface. Researchers have rarely looked at these biospecimens in large groups of healthy persons. It is conceivable that we may discover protective genes, cellular proteins, or intestinal flora that reduce one's susceptibility to a host of autoimmune disorders, for example. The possibilities are almost limitless.

Even more innovative may be the insights gleaned from collecting geospatial and environmental data such as weather, air quality, environmental pollutant levels, food deserts, walkability, population density, and climate change. Once upon a time, the suggestion that air pollutants contributed to emphysema and other pulmonary disorders was pure speculation. New associations between some of the social determinants of health—including a lack of fresh fruits and vegetables in local grocery stores or neighborhoods that are too dangerous to walk in—and type 2 diabetes are not far-fetched. But they do require substantive scientific evidence before they are taken seriously by national and state policymakers and by clinicians in everyday practice.

Finally, the PMI Working Group plans to collect social networking data, including Twitter feeds, social contacts from cellphone text and voice, and even OTC medication purchases. Wading through these mountains of data and determining what to collect and analyze is an enormous task in itself, and whether it is worthwhile remains to be seen.

THE ROLE OF ELECTRONIC HEALTH RECORD DATA

EHRs are a rich source of health data, which if analyzed in creative and imaginative ways, will likely reveal relationships between risk factors and diseases and between a variety of behaviors and optimal health. To take full advantage of these relationships, the PMI working group has made several recommendations. The PMI cohort should develop standardized and, to the extent possible, automated mechanisms to acquire clinical data efficiently from participating health-care provider organizations. It also needs to develop and implement a rigorous data curation program for the core datasets to create analysis-ready datasets for a broad range of uses.

Additional guidelines for the use of EHR data are as follows:

- Periodically implement and revise phenotype algorithms for a number of core diseases and outcomes of interest.
- Develop central resources to support data curation and implementation of algorithms to identify common phenotypes of interest.
- Support development and adoption of automated text analytics that can be used both centrally and locally to extract for research purposes the information contained in narrative clinical documents.

Unfortunately currently available EHR products are not fully equipped to handle some of the most important data elements now being generated by PMIs. A case in point is the genomic data that will serve as one of the key foundation stones for individualized patient care. Most EHRs do not have dedicated sections to record single nucleotide polymorphisms (SNPs) or other specific genomic data points. More importantly, most physicians who do not specialize in genetic medicine are uncertain how to interpret the data or use it in clinical practice.

Similarly, EHRs do not provide adequate space for all the psychosocial parameters that the PMI hopes to measure. Clearly, today's EHR systems have strengths and weaknesses and were not designed to take on a project of this size. We will explore the role of EHRs in precision medicine in more detail in Chapter 5.

PRECISION MEDICINE AT HARVARD

The federally sponsored PMI certainly is not the only authoritative source of information on personalized medicine, nor the only research initiative that aims to move the specialty forward. Among the thought leaders in the field are the technologists, data scientists, and clinicians from Harvard University's department of biomedical informatics and Partners HealthCare Personalized Medicine [15,16].

Both the PMI project and the Harvard initiatives have emphasized one critical shortcoming in medicine as it is practiced today. That shortcoming is summarized on the opening page of the biomedical informatics department's website:

> Over 100 years ago, Abraham Flexner reviewed the state of U.S. medical education and its impact on the practice of medicine. What he reported about the lack of modern science in medicine caused half of the medical schools in this country to close. Today we are at another Flexnerian moment. Medicine and biomedical science are data- and knowledge-processing enterprises that are largely conducted without any of the modern tools of quantitative analysis or automation…. It is our mission to ensure that the next 100 years result in the breakthrough treatments and scientific insights we presaged in a New England Journal of Medicine piece entitled 'A Glimpse of the Next 100 Years in Medicine.'

Critics of the precision medicine movement often complain that the likelihood of obtaining any clinically meaningful information from this approach is remote, and the return on investment is decades away. It would be far wiser to concentrate on public health initiatives and traditional clinical research. But it is clear from the above statement that Harvard's biomedical informatics department does not share that devotion to the status quo.

Issac Samuel (Zak) Kohane, MD, PhD, chair of the department, encourages others to "go rogue," break rules, rattle cages, and bust barriers. There are many ways to express this kind of thinking. An Apple commercial once expressed it in these terms: "Here's to the Crazy ones. The misfits, the rebels, the troublemakers, the round pegs in square holes. The ones who see things differently… Because the people who are crazy enough to think they can change the world are the ones who DO."

Applying new technology to health care demands this kind of unconventional thinking and the courage to transform medicine with disruptive solutions to age-old problems.

The Precision Medicine 2016: Rogue Therapeutics Conference illustrates this transformative mind-set. Held at Harvard Medical School, the conference focused on creative clinicians, engineers, technologists, and patients who have been unwilling to wait for conventional medicine to catch up to their progressive thinking.

For example, the conference included a presentation about the development of a bionic pancreas, which its creator Edward Damiano, PhD, refers to as "the epitome of personalized medicine in your pocket." The one-size-fits-all approach to type 1 diabetes utilizes conventional insulin pumps that require constant glucose monitoring and adjustments in insulin doses, diet, and exercise to maintain blood glucose levels within a normal window. The new device is capable of mimicking the human pancreas, which responds automatically to a variety of physiological changes by secreting varying doses of insulin and glucagon to keep blood glucose steady. The bionic pancreas will begin the final stage of human clinical trials in 2017 in the hope of gaining FDA approval [17].

Harvard's biomedical informatics department is also devoting much of its resources to research on undiagnosed and rare diseases. Since many of these disorders affect a universe of one, they are the perfect breeding ground for innovations in personalized medicine. For instance, the Rogue Therapeutics Conference featured the work that the biomedical informatics department is doing in conjunction with NIH Undiagnosed Disease Program, which is part of the National Human Genome Research Institute.

William Gahl, director of the Undiagnosed Disease Program, pointed out during his conference presentation that people with undiagnosed diseases are relatively abandoned by family, friends, and physicians alike. About 4000 patients have submitted their medical records to the program seeking assistance, and about 1000 have been accepted into the program. Accepted patients are invited to the NIH Clinical Center for a free 1 week inpatient stay for testing and analysis. The program has uncovered 100–200 unique diseases for which the program was able to establish a detailed phenotype and conduct gene sequencing on the patient and their families. Since the inception of the program, NIH has invested additional funds to expand it to six additional clinical sites around the country. There is also international expansion taking place to create a network of similar programs for the study of undiagnosed diseases, which include Australia, Austria, Italy, and Japan. The program has also piqued the interest of pharmaceutical firms seeking to create therapeutic agents for these new disorders.

PRECISION MEDICINE AT COLUMBIA UNIVERSITY

Like Harvard University, Columbia is deeply committed to research in genomics and data science. Tom Maniatis, PhD, is director of Columbia's university-wide PMI, which also encompasses proteomics, bioinformatics, systems biology, data and computational science, as well as core science, engineering, and other disciplines. Columbia's clinicians are already using genomic analysis to sequence the DNA of patients' tumors to determine which FDA-approved drugs will most likely target an individual's cancer.

Unfortunately, this state-of-the-art technology has little application outside of academic centers. During a recent Columbia University conference on how precision medicine is shaping cancer research, it became obvious that most adult patients being cared for by community oncologists have yet to receive this kind of targeted personalized therapy. They still have to rely on standard chemotherapy and radiation, a so-called "scorched body approach" to medicine.

In addition to pursuing its own PMIs, Columbia University will be deeply involved in the federally sponsored PMI. NIH has invested heavily in a coalition of health-care organizations to help it launch the project, including Vanderbilt University, Verily Life Sciences, and the Broad Institute. Columbia will help curate data from EHRs, medical and pharmaceutical databases, and payer databases. It will also standardize the information, ensure data quality, and convert the data into a usable format. The university is well qualified to take on this responsibility in light of the fact that it has been managing massive amounts of health-care data through the Observational Health Data Sciences and Informatics initiative, a combined repository of more than 600 million patient records in 14 countries [18].

JOHNS HOPKINS APPROACH TO INDIVIDUALIZED CARE

Johns Hopkins University prefers to call its precision medicine program an Individualized Health Initiative [19]. The program is involved in several projects that are helping to target subgroups and individuals and to reduce the risk of adverse drugs reactions among patients.

Among its *in*Health projects are several focused on autoimmune disease, cardiology, and cancer. Recognizing that autoimmune diseases are diverse and progress at different rates in different patients, the *in*ADM group has started studying Johns Hopkins patients with scleroderma, which causes hardening of the skin and several organ complications that vary widely in their severity among individual patients. They have developed a computational approach that has helped to define new scleroderma subtypes and are analyzing tissue samples to determine if there are immune system biomarkers that will identify those unique subtypes. The next step is to use their findings to "model and predict the individual disease trajectory of scleroderma patients and adapt this model to other autoimmune diseases, including rheumatoid arthritis and lupus" [20].

The university's *in*Car group is also bringing personalized medicine to interventional cardiology. The group points out that while such procedures can be lifesaving, the process by which patients are chosen for intervention is imprecise. To better match individual patients to specific procedures, Johns Hopkins cardiologists have created a learning health system that integrates several sources of patient data into a database that is routinely analyzed to determine the optimal treatment strategy. The learning health system is designed to accomplish three goals:

- Continuously update scientific knowledge on the benefits of interventional cardiovascular procedures,
- Define which patients are most likely to benefit from which procedures, and
- Provide decision support tools for cardiologists to use with their patients to inform the best course of action.

One of the most innovative programs launched at Johns Hopkins has been labeled The Learning Methodologies Core. Its purpose is to take the clinical research process into the future by developing new statistical methods, decision-making tools, and research designs, including the adaptive trial design. With the help of this new initiative, the university is making advances in the individualized management of pneumonia and prostate cancer. In the program, Pneumonia Etiology Research for Child Health (PERCH) program, researchers are creating statistical tools that combine rest results from several diagnostic procedures to predict which pathogen is most likely to be the causative agent in an individual patient. Since pneumonia can be caused by more than 30 distinct pathogens and most diagnostic tests used in clinical medicine fail to precisely pinpoint that pathogen, the PERCH program is certainly a step in the right direction.

The diagnosis and treatment of prostate cancer has become a source of controversy in the medical community and the popular press since the evidence to support the value of PSA has been called into question. Studies indicate that many men have been needlessly treated for slow growing malignancies and must now cope with the impotence and urinary incontinence that often accompany treatment. Learning Methodologies Core investigators are devising models that can estimate the effects of various treatment strategies on clinical outcomes, while at the same time factoring in patients' preferences and the severity of each patient's disease. "The end–product will be a support tool - designed for clinician and patient use – that will facilitate treatment decisions based on individualized risk prediction" [21].

Johns Hopkins researchers have also been able to add genetic risk scores to a patient's family history to help more precisely pinpoint a man's risk of prostate cancer. Chen et al. have used SNPs associated with the risk of prostate cancer to supplement the family history of the disease to identify a subgroup of patients at higher risk. When family history alone was used, 17% of men were found to have a positive history, and the cancer detection rate was 29.0%. Among men with no family history of the disease, 23.4% developed the malignancy. However, when a positive family history was combined with a genetic risk score above 1.4, based on the presence of the genetic variants, more than twice as many men were classified as high risk, and the cancer detection rate increased to 31%, compared to 20.6% in those without the genetic risk score [22].

MAYO CLINIC CENTER FOR INDIVIDUALIZED MEDICINE

The Mayo Clinic is pursuing a wide variety of precision medicine projects in several specific areas: biomarker discovery, the microbiome, pharmacogenomics, epigenomics, and oncology. Investigations on the role of the microbiome, which includes the intestinal bacteria in the human gut, are looking into composition of the intestinal microflora of patients with gluten sensitivity. Researchers are also trying to characterize the fecal composition of patients with *Clostridium difficile* infection to develop tools that can predict which individuals will respond to treatment and who is at risk for the disease. They are also studying the possible role of intestinal microbes in the etiology of rheumatoid arthritis in the hope of developing predictive and diagnostic tools for individuals at high risk for the disease. Similar projects are underway to determine the role of the microbiome in colon cancer and bacterial vaginosis.

The Clinic has also embarked on numerous clinical trials that have an individualized medicine component. One seeks to identify the environmental and genetic risk factors for pediatric multiple sclerosis, another is a pilot study using proteomic and genomic profiling for patients with metastatic breast cancer, and a third is investigating the genomics of primary biliary cirrhosis and primary sclerosing cholangitis.

The Mayo Clinic has also been comparing the diagnostic value of genetic testing to genomic testing in the hope of detecting DNA variants in individual patients that have pathological implications. Clinical medicine is slowly transitioning from the traditional emphasis on testing for specific genes, e.g., the gene for Huntington's disease, to an approach that casts a wider net, like whole exome sequencing.[1] To examine the value of the latter, Lazaridis and colleagues from Mayo studied the outcomes of patients who had genome sequencing at their individualized medicine clinic [23].

During an 18 month period, the Individualized Medicine Clinic received 82 requests from clinicians, and 51 patients submitted specimens for testing. Whole exome testing was able to detect 15 disorders, a positive diagnostic yield of 29%, with an average cost per patient of about $8000. Lazaridis et al. concluded "The significant diagnostic yield, moderate cost, and notable health marketplace acceptance for WES compared with conventional genetic testing make the former method a rational diagnostic approach for patients on a diagnostic odyssey."

STANFORD UNIVERSITY EMPHASIZES PATIENT ENGAGEMENT

With all the emphasis on genomic analysis, biomarkers, proteomics, and the like, it is easy to overlook the most important ingredient in personalized medicine—the individual receiving the care—who is much more than a collection of genes, metabolites, and lab values. Without an individual's engagement in his or her care, these other measures are far less effective. And while many individuals are not interested in participating in their own care, those who are can benefit greatly from a health-care team eager to work with them as partners rather than passive recipients. Few precision

---

[1] The genome, which is the entire DNA complement of an individual, is composed of exons and introns. The exons are sections of the genome that are transcribed into RNA, which in turn is responsible for the synthesis of body proteins. Introns are DNA segments in between the sections of DNA that comprise the exome. Introns, or intragenic regions, are sometimes referred to as "junk DNA" although that is not an accurate term. Exome sequencing is a practical relatively inexpensive way to search for known gene disorders.

medicine programs have devoted as much energy to patient engagement, personal responsibility, and mutual respect between clinicians and patients as Stanford University.

To foster the bonds between providers and patients, Stanford University has created the "Everyone Included" program. In the words of the management team:

> Everyone Included™ is a framework for healthcare innovation, implementation and transformation based on principles of mutual respect and inclusivity. It is the culmination of six years of co-creation with patients, caregivers, providers, technologists, and researchers at Stanford Medicine X that has resulted in a series of design and leadership principles intended to drive collaborative healthcare innovation efforts. Our work has been field tested and iteratively improved over the past six years at our Stanford programs and convenings worldwide.

The program is grounded in 10 design principles, the aim of which is to create innovative solutions to once intractable health problems. Like the Harvard Rogue Therapeutics Conference mentioned above, the first principle is "Be a rebel," i.e., stand up for what you believe health care should be. In other words, do not be content with the status quo. The remaining principles include several obvious but often overlooked themes in routine patient care: value each person, be human, be human-centered, codesign, facilitate connections, treat with dignity, and provide a stage from which the hardest, most important stories may be told [24].

Of course, Stanford's PMI is not just about patient engagement. Bioinformatics, mobile technology, genomics, and data science are also priorities. For example, Stanford Medicine is working with the American Heart Association to help individualize cardiac care. It has created the MyHeart Counts app, which is built on Apple's ResearchKit platform to allow patients to share their physical activity and cardiac risk factor data with researchers. The goal is to help determine the complex relationship between heart health and behavior.

Research conducted at Stanford University School of Medicine has also found a technique that may help identify individuals who are more likely to respond poorly to medication known to cause cardiac damage. Joseph Wu, MD, PhD, and his colleagues have created specialized cardiomyocytes (heart muscle cells) from stem cells from individual volunteers and tested how they respond to rosiglitazone, an antidiabetic agent, and tacrolimus, an immunosuppressant used to inhibit organ rejection after transplantation. Both drugs have a tendency to damage heart tissue. The stem cell generated heart muscle cells served as a biomarker to identify individuals who do not rosiglitazone tolerate well. According to Dr. Wu: "…the patient derived iPS [induced pluripotent stem] cell platform gives us a surrogate window into the body and allows us to not only predict the body's function but also learn more about key disease-associated pathways" [25].

Our sampling of precision medicine programs and initiative is by no means exhaustive. Other programs that illustrate the interest in this new field include the Penn Center for Precision Medicine, the University of Pittsburgh Medical Center, Institute of Personalized Medicine, and Genomics Data Commons, which is located at the University of Chicago.

REFERENCES

[1] Genomics England. Genomics England's response to President Obama's precision medicine initiative. January 30, 2015. https://www.genomicsengland.co.uk/genomics-englands-response-to-president-obamas-precision-medicine-initiative/.

[2] Kichko K, Marschall P, Flessa S. Personalized medicine in the U.S. and Germany: awareness, acceptance, use and preconditions for the wide implementation into the medical standard. J Pers Med 2016;6(2):15. http://www.mdpi.com/2075-4426/6/2/15/htm.

[3] Next Big Future. China's $9.2 billion precision medicine initiative could see about 100 million whole human genomes sequenced by 2030 and more if sequencing costs drop. June 7, 2016. http://www.nextbigfuture.com/2016/06/chinas-92-billion-precision-medicine.html.

[4] The precision medicine initiative cohort program – building a research foundation for 21st century medicine: precision medicine initiative (PMI) working group report to the advisory committee to the director. NIH; September 17, 2015. http://acd.od.nih.gov/reports/DRAFT-PMI-WG-Report-9-11-2015-508.pdf.

[5] The White House. FACT SHEET: President Obama's precision medicine initiative. January 30, 2015. https://www.whitehouse.gov/the-press-office/2015/01/30/fact-sheet-president-obama-s-precision-medicine-initiative.

[6] Pear R. Uncle Sam wants you — or at least your genetic and lifestyle information. New York Times; July 23, 2016. http://www.nytimes.com/2016/07/24/us/politics/precision-medicine-initiative-volunteers.html?ref=topics&_r=3.

[7] Wazwaz N. It's official: the U.S. Is becoming a minority-majority nation. U.S. News and World Report; July 6, 2015. http://www.usnews.com/news/articles/2015/07/06/its-official-the-us-is-becoming-a-minority-majority-nation.

[8] Halbert CH, McDonald J, Vadaparampil S, Rice L, Jefferson M. Conducting precision medicine research with African Americans. PLoS One 2016;11(7):e0154850. http://dx.doi.org/10.1371/journal.pone.0154850. http://journals.plos.org/plosone/article?id=10.1371%2Fjournal.pone.0154850.

[9] U.S. Preventive Services Task Force. Prostate cancer screening. May 2012. http://www.uspreventiveservicestaskforce.org/Page/Document/UpdateSummaryFinal/prostate-cancer-screening.

[10] Vargas A, Harris CC. Biomarker development in the precision medicine era: lung cancer as a case study. Nat Rev Cancer 2016;16:525–37.

[11] Giancardo L, Sánchez-Ferro A, Butterworth I, Mendoza CS, Hooker JM. Psychomotor impairment detection via finger interactions with a computer keyboard during natural typing. Sci Rep 2015;5:9678.

[12] Go AS, Mozaffarian D, Roger VL, et al. Executive summary: heart disease and stroke statistics–2014 update: a report from the American Heart Association. Circulation 2014;129(3):399–410.

[13] Comstock J. NIH awards $120M to scripps, others, to enroll 350K participants in precision medicine initiative via mobile apps. MobiHealthNews; July 7, 2016. http://mobihealthnews.com/content/nih-awards-120m-scripps-others-enroll-350k-partici-pants-precision-medicine-initiative-mobile.

[14] National Institutes of Health. Participant technologies center. July 6, 2016. https://www.nih.gov/precision-medicine-initiative-cohort-program/participant-technologies-center.

[15] Partners Healthcare Personalized Medicine. http://personalizedmedicine.partners.org/.

[16] Harvard Medical School, Department of Biomedical Informatics. https://dbmi.hms.harvard.edu/.

[17] Rimer S. Hope for the battle against type 1 diabetes: benefit corporation founded by parents of children with the disease. BU Today; April 4, 2016. http://www.bu.edu/today/2016/beta-bionics-artificial-pancreas/.

[18] Columbia University Medical Center. Precision medicine: a Columbia University initiative: precision medicine encompasses all the right reasons for a new approach to health cares. March 26, 2015. http://newsroom.cumc.columbia.edu/precision-medicine/.

[19] Johns Hopkins individualized health initiative. 2015–2017. http://hopkinsinhealth.jhu.edu/.

[20] Johns Hopkins individualized health initiative. Hopkins program to individualize auto-immune disease management (inADM). 2015–2017. http://hopkinsinhealth.jhu.edu/our-research/our-programs/inadm.

[21] Johns Hopkins learning methodologies core. 2015–2017. http://hopkinsinhealth.jhu.edu/our-research/our-programs/learning-methodologies-core.

[22] Chen H, Liu X, Brendler CB, et al. Adding genetic risk score to family history identi-fies twice as many high-risk men for prostate cancer: results from the prostate cancer prevention trial. Prostate 2016;76(12):1120–9.

[23] Lazaridis KN, Schahl KA, Cousin MA, et al. Outcome of whole exome sequencing for diagnostic odyssey cases of an individualized medicine clinic: the Mayo clinic experi-ence. Mayo Clin Proc 2016;91(3):297–307.

[24] Stanford Medicine X: Everyone Included. http://www.everyoneincluded.org/.

[25] Conger K. Heart muscle made from stem cells aids precision cardiovascular medi-cine. Stanford Medicine. August 18, 2016. http://med.stanford.edu/news/all-news/2016/08/ips-cell-derived-heart-cells-predict-drug-toxicity.html.

CHAPTER THREE

The Role of Genomics

Clinical genomics may not be synonymous with precision medicine, but it is certainly one of the key components that are helping practitioners realize its promise. Among the specialties most impacted by genomics: oncology and pharmacology.

Since genomics is a relatively new discipline and many clinicians and technologists are unfamiliar with the science behind its clinical application, let's review the basics first.

GENOMIC BASICS

Box 1.1 in Chapter 1 (Pages 19–20) explained the "ABCs" of DNA, including a brief explanation of the structure of the chromosome, the DNA molecule, base pairs, RNA, transcription, translation, single nucleotide polymorphism (SNP), and how amino acids, which are the building blocks of all the body's proteins, are derived from the encoding process.

Our understanding of these basic principles has helped researchers and clinicians address numerous genetic disorders that have plagued humanity for centuries. We now have laboratory testing procedures to identify a wide variety of chromosomal or single-gene diseases such as Down syndrome, cystic fibrosis, sickle cell disease, and muscular dystrophy. But the era of genetic medicine is slowly being supplemented by genomic medicine, which analyzes the entire genetic makeup of an individual to detect several genes that may be contributing in more subtle ways to disease and that are interacting with environmental triggers that turn genetic predisposition into clinical reality. A review paper in the *New England Journal of Medicine* points out the following: "As a result of genomic discoveries, increasing numbers of clinical guidelines now suggest incorporating genomic tests or therapeutics into routine care…. Regardless of where medicine is practiced, genomics is inexorably changing our understanding of the biology of nearly all medical conditions. How can any clinician understand the diagnosis

Realizing the Promise of Precision Medicine
ISBN 978-0-12-811635-7
http://dx.doi.org/10.1016/B978-0-12-811635-7.00003-8

and treatment of breast cancer… without a rudimentary understanding of genomic medicine?" [1].

One of the revelations that have resulted from the Human Genome Project is that there is no normal human genome. We are all mutants. This is the case despite the fact that approximately 99.6% of the base pairs that exist within our genomes are identical, leaving only 0.4% to provide the variability responsible for human individuality. With rare exceptions, all humans have 46 chromosomes within the nucleus of each of their cells, with the exception of germ line cells—namely sperm and eggs—which each contain 23 chromosomes. Approximately 25,000–30,000 genes are distributed among our chromosomes. And within nongerm line cells, about 6 billion base pairs reside in these genes. That small 0.4% variation in base pairs translates into about 24 million differences between any two individuals. And since the sequence of base pairs within our genes is ultimately responsible for numerous enzymes and other body proteins, this "small" variation is largely responsible for our biochemical individuality.

How does genetic individuality inform our understanding of health and disease? An individual's susceptibility to disease is determined by two types of mutations. Genetic defects that occur in one's germ cells are derived from a person's descendants and are transmitted through sperm and egg. Mutations in the genes of a one's nongerm cells—called somatic cells—are not passed on to offspring but have been implicated in the development of cancer and a variety of other disorders.

Testing for individual mutations has long been a part of clinical medicine. Cell-free fetal DNA testing, for instance, is recommended to test for the presence of Down syndrome in women whose ultrasound results suggest the presence of the chromosomal disorder. Similarly, patients suspected of harboring the gene that causes hereditary hemochromatosis, an iron storage disease, can be tested for the presence of specific gene mutations to confirm the diagnosis.

But genomic testing takes genetic medicine to a whole new level. Rather than testing for an individual mutation, it can analyze the entire genome to detect millions of genetic variants. This sophisticated analysis can be accomplished using whole genome sequencing, genome-wide association studies or GWASs, and computer chip technology, assisted by fluorescence detection.

The GWAS analyzes the genomes of a large group of individuals, comparing DNA sequences from persons with a specific disease to the DNA of a control group of healthy persons. The process initially involves collecting blood samples or rubbing a cotton swab inside the mouth to collect cells that can be analyzed. Once the specimens are purified, the DNA is placed in small gene chips (also referred to as microarrays) and then scanned by automated laboratory devices. The machines search for SNPs. Finally, researchers compare the SNPs from diseased subjects to the SNPs of normal persons to detect genetic variants that may be associated with the disease.

The Framingham Genetic Research Study is one of many examples of a GWAS. Launched in 2006, it has been collecting data to help identify genes that may underlie cardiovascular and other chronic disorders. It involves up to 500,000 genetic analyses of DNA from 9000 participants in the famous Framingham Heart Study (FHS). Details on the genomic information collected in that study are available on the Genetic Data page of the FHS website [2].

Three GWASs have identified a gene variant for complement factor H that has been linked to a common type of blindness. Complement factor H is responsible for the production of a protein that regulates inflammation. That clue has pointed investigators in a new direction because inflammation had not previously been considered as a contributing factor for this type of blindness [3]. GWASs have also detected SNPs that may increase the risk of type 2 diabetes, Parkinson's disease, heart disorders, obesity, Crohn's disease, and prostate cancer. However, it is important to keep in mind that GWASs only establish associations between mutations and disease, not cause and effect relationships.

One of the technologies that have made genome analysis faster and less expensive is the DNA microarray or gene chip. These chips can be used for a variety of purposes: The gene chips are valuable in diagnostic testing. Some academic cancer centers now use these tools for patients to determine what pathogenic mutations exist with tumor cells. The technology has also been used to find out which genes are active and which have been turned off in tissues. In that case, technicians will want to isolate RNA in a tissue sample rather than DNA. (Since microarray analysis is used for a variety of clinical and research applications, the explanation that follows will not apply to every conceivable application.)

Microarray analysis incorporates the use of a chip that resembles a microscope slide or microprocessor chip. The chip is composed of numerous slots or wells, in which samples are placed. Let's assume we want to analyze tumor cells from a patient with colon cancer. A sample of the malignant tissue and a sample of normal tissue are both denatured to separate the two complementary strands of DNA. (Since DNA is a twisted ladder or helix, it is composed

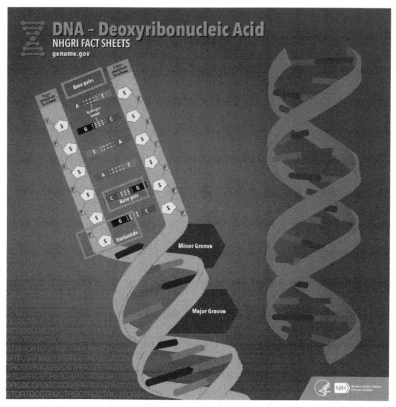

Figure 3.1   DNA consists of a twisted ladder or helix composed of 2 "side rails" with each rail connected to one of four bases: A adenine, T, thymine, C, cytosine, and G, guanine. A and T always pair with one another. i.e., they are complementary; the same applies to C and G.

of 2 "side rails" with each rail connected to one of four bases: A adenine, T, thymine, C, cytosine, and G, guanine; A and T always pair with one another. i.e., they are complementary; the same applies to C and G.) (See Figs. 3.1 and 3.2).

Then these long DNA strands are cut into smaller fragments and labeled with separate fluorescent dye. The patient's sample will glow green, and the control will glow red. These samples are then added to the gene chip, which has been seeded with preassigned DNA sequences. If the patient has a mutation, his or her DNA will not bind properly to certain chemically synthesized DNA probes in the chip, which represents the normal sequence, but will bind to the fragment on the chip that represents the mutated DNA. DNA microarray analysis should be differentiated from whole genome sequencing, which involves a different technology. In the latter, the goal is to determine the precise order of all the nucleotides within a person's genes.

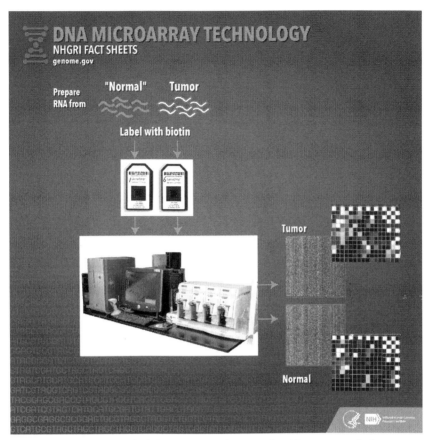

Figure 3.2 Microarray analysis incorporates the use of a chip composed of numerous slots or wells, in which samples are placed. A sample of the malignant tissue and a sample of normal tissue are both denatured to separate the two complementary strands of DNA and then compared.

PUTTING GENOMICS TO USE IN PRECISION MEDICINE

There are numerous success stories to illustrate the value of genomics in individualizing medical care. It has proven invaluable in providing definitive diagnoses in severe intellectual disability, for example. In the Netherlands, exome sequencing of 100 individuals with severe mental disabilities whose parents were not affected revealed genes and mutations that were contributing to the syndromes. Exome sequencing, unlike whole genome sequencing, only analyzes the coding regions of a person's DNA, ignoring the introns or noncoding areas of their DNA. When de Ligh et al.

sequenced the exons of more than 21,000 genes in 100 persons with IQs below 50, they were able to identify 79 new mutations in 53 of the 100 subjects, several of which were likely contributing to the individuals' intellectual disorders [4].

Gene sequencing for hereditary hypertrophic cardiomyopathies has also proven useful in diagnosing and treating this family of cardiac disorders. Partners Healthcare Personalized Medicine group provides extensive genomic analysis to help detect these cardiomyopathies. The most common forms are hypertrophic cardiomyopathy (HCM) and dilated cardiomyopathy (DCM), followed by arrhythmogenic cardiomyopathy and left ventricular non-compaction. The Pan Cardiomyopathy test panel used by Partners has a detection rate of approximately 35% for HCM, ~37% for DCM, and ~50% for arrhythmogenic cardiomyopathy [5]. And as we discussed in Chapter 1, there is a new drug available to target one of the underlying causes of the disease, called MYK-461.

One of the challenges that clinicians and patients face when dealing with relatively rare disorders that would benefit from gene sequencing is third party reimbursement. Many insurers have not kept up with the latest advances in genomics or have yet to be convinced that such testing is cost-effective. Genetic testing for diagnosis and management of hereditary cardiomyopathies, however, is one of the areas in which coverage is available. Empire Blue Cross Blue Shield, for instance, has concluded that genetic testing for hereditary HCMs is medically necessary, assuming certain criteria are met [6].

Patients with certain types of colorectal cancer can also receive more personalized treatment if they are tested for a mutation that makes them responsive to aspirin therapy. Aspirin inhibits the activity of phosphatidylinositol 3-kinase (PI3K), which is part of a key signaling pathway. Fifteen to 20% of patients have a PIK3CA mutation in the gene that is responsible for the synthesis of a protein that is part of the PI3K pathway. By interfering with this biochemical process, aspirin therapy has been found to increase survival of colorectal cancer patients with the mutation, when compared to patients without the gene mutation. The Dana–Farber Cancer Institute investigators who discovered the link between this mutation, aspirin therapy, and better survival concluded their report by saying "Regular use of aspirin after diagnosis was associated with longer survival among patients with mutated-PIK3CA colorectal cancer, but not among patients with wild-type PIK3CA cancer. The findings from this molecular pathological epidemiology study suggest that the PIK3CA mutation in colorectal cancer may serve as a predictive molecular biomarker for adjuvant aspirin therapy" [7].

Pediatricians have known for decades that newborn genetic testing can help detect a variety of severe disorders. In fact, in the United States, between 29 and 50 severe, inherited treatable genetic diseases are screened for through public health programs. But there are *many* more genetic disorders worth considering, and scanning for the relevant mutations would allow clinicians to diagnose and treat them before they produced irreversible damage. In fact, there are almost 600 recessive genetic disorders in existence that have pediatric presentations that correspond to phenotypes and genes that have been shown to cause symptoms. Some researchers believe whole genome sequencing can meet that need [8].

Similarly both pediatric and adult patients may eventually benefit from DNA testing that can detect the likelihood of a heart transplant recipient rejecting the organ. The blood test detects cell-free circulating DNA from the donor's organ in the blood of the patient who received the heart. The test is superior to the current method of determining tissue rejection, which involves biopsy. The DNA signature can be detected in a patient's blood up to 5 months before it is discovered by biopsy [9].

ENTERING THE ERA OF PRECISION ONCOLOGY

No doubt cancer patients would like to see oncology advance to the point where every malignancy can be precisely mapped genetically, and to see those genetic blueprints provide clinicians with personalized treatment protocols to eradicate their tumor. Unfortunately, we are not close to that kind of individualized diagnosis and treatment. But that is not to suggest that genomics has provided no actionable insights that have a direct impact on the prevention and treatment of cancer.

Mutations in BRCA1 and BRCA2 genes greatly increase the risk of breast cancer and ovarian cancer. Between 3% and 5% of breast cancers are hereditary, and most can be traced back to genetic defects in these two genes. Of course, that also means that 95%–97% of women with breast cancer have developed a form of the disease unrelated to either of these two mutations. Two recent large-scale studies may help to individualize preventive strategies for women with BRCA mutations and for those without.

Rebbeck et al. [10] analyzed BRCA1 and BRCA2 mutations in more than 31,000 women from 33 countries on 6 continents. Not content to simply confirm that these mutations put women at risk, they looked more closely at the specific mutations in these two genes to see if that risk varied in different subgroups. They found that different BRCA1/2 mutations are associated with significantly different levels of risk of breast and ovarian cancer depending on where the mutations occur within the genes. If a mutation was located near the ends of the BRCA1 coding sequence, women were more likely to develop breast cancer than if the mutations were located elsewhere. Similarly, women had a greater risk of developing ovarian cancer if the BRCA1 mutation occurred in the middle section of the coding sequence.

The researchers labeled various mutational "bins" on the BRCA1 DNA sequence from c.1 to c.5563. The mutations were grouped by type and function, including frameshift, nonsense, missense, splice site, and by in-frame or out-of-frame status. (See Box 3.1 for definitions.) As a result of their analysis, they were able to identify mutations from c.179 to c.505 that were breast cancer cluster regions; a similar pathogenic cluster was identified

Box 3.1 Types of Mutations

As we mentioned earlier, there is no single "normal" genome. We are all mutants in the sense that all humans have some DNA sequences that vary from the average. But not all mutations are detrimental. Those that are, sometimes referred to as pathogenic mutations, are typically responsible for the production of abnormal proteins or the absence of these proteins, which in turn disrupts a wide range of physiological activities. What follows is an explanation of a few important mutations that can cause or contribute to disease, derived from the NIH National Human Genome Institute and other sources.

A *frameshift mutation* is a type of mutation involving the insertion or deletion of a nucleotide in which the number of deleted base pairs is not divisible by three. "Divisible by three" is important because the cell reads a gene in groups of three bases. Each group of three bases corresponds to one of 20 different amino acids used to build a protein. If a mutation disrupts this reading frame, then the entire DNA sequence following the mutation will be read incorrectly.

A *nonsense mutation* is the substitution of a single base pair that leads to the appearance of a stop codon where previously there was a codon specifying an amino acid. The presence of this premature stop codon results in the production of a shortened, and likely nonfunctional, protein.

A *missense mutation* is when the change of a single base pair causes the substitution of a different amino acid in the resulting protein. This amino acid substitution may have no effect, or it may render the protein nonfunctional.

A *deletion mutation* is a type of mutation involving the loss of genetic material. It can be small, involving a single missing DNA base pair, or large, involving a piece of a chromosome.

at c.5261–c.5563, in other words, at the 2 ends of the BRCA1 coding sequence. Ovarian cancer cluster regions were discovered from c.1380 to c.4062, i.e., in the middle of the DNA strand. Similarly cancer-related DNA regions were identified along the BRCA2 gene.

Rebbeck et al. conclude their analysis: "Breast and ovarian cancer risks varied by type and location of BRCA1/2 mutations. With appropriate validation, these data may have implications for risk assessment and cancer prevention decision making for carriers of BRCA1 and BRCA2 mutations."

The clinical implications of a recent analysis by Paige Maas and associates may be more widespread because it applies to a much larger population than women with BRCA mutations [11]. They examined more than 33,000 breast cancer cases and 33,000 controls to analyze the value of a collection of SNPs—variants that did not include BRCA1 or BRCA2 mutations. These 77 mutations comprised a polygenic risk score that investigators suspected might allow clinicians to stratify women's risk of developing breast cancer. As you might expect, the analysis revealed that women who had the highest genetic risk score were most at risk. For example, women in the highest 1% risk score group were more than three times as likely to develop breast cancer as women in the middle score range [12]. Put another way, "the lifetime risk of breast cancer for women in the lowest and highest quintiles of the risk score were 5.2% and 16.6% for a woman without family history, and 8.6% and 24.4% for a woman with a first-degree family history of breast cancer."

While these statistics may allow clinicians to provide more individualized advice to patients about the likelihood of developing breast cancer, they can also help clinicians individualize advice on the value of various modifiable risk factors. Maas et al. found that women most likely to develop breast cancer because of nonmodifiable risk factors—including the SNPs, a family history of the disease, and height—had a lower risk of the disease if they had a low body mass index, did not smoke or drink, and did not use menopausal hormone therapy—all factors that individuals have more control over. In fact, their risk was comparable to that of the average woman in the general population.

Advances are also being made in the application of genomics to the treatment of breast cancer. F. Cardoso, from the Champalimaud Clinical Center–Champalimaud Foundation, in Lisbon, Portugal, and associates have been able to demonstrate that a 70 gene test panel called MammaPrint can help clinicians and patients with early stage breast cancer and make a more informed choice about whether or not to undergo adjuvant chemotherapy [13].

The test panel uses microarray technology to analyze the genes in an individual patient's tumor to determine how active certain genes are. That in turn helps estimate who is at risk for metastasis and thus in greater need of adjuvant chemotherapy.

Cardoso et al. enrolled over 6600 women with early stage breast cancer and performed two types of risk assessments on each of them. The clinical risk assessment looked at patients' estrogen receptor positivity or negativity, HER2 status, the number of lymph nodes affected, the size of the tumor, and its grade. The genomic analysis used the MammaPrint test panel to arrive at a genetic risk score. Women with low scores on both assessments were not given chemotherapy, and those at high clinical and high genomic risk were given chemotherapy.

About 23% of patients presented with high clinical risk but low genomic risk. Among those patients in this group who did not receive chemotherapy, 5-year survival without distant metastasis was 94.7%. That was only 1.5% points less favorable than the survival observed in the patients with high clinical risk and low genomic risk who *were* given chemotherapy. That suggests that many individuals who seem to need adjuvant chemotherapy based on their pathological and clinical findings probably do not need the therapy and can be spared the toxicity that accompanies the treatment. And while tests such as MammaPrint do not allow clinicians to precisely identify a single individual's need for chemotherapy, they do contribute to personalized medicine by establishing the existence of subgroups who are more likely to benefit from the treatment.

In Chapter 2, we discussed the value of biomarkers in precision medicine. Some of these biomarkers take advantage of discoveries about the genetics of cancer. The presence of epidermal growth factor receptor (EGFR) as a biomarker has been shown to help distinguish patients with lung adenocarcinoma from one another. For example, the subgroup of patients with EGFR-positive nonsmall cell lung adenocarcinoma can now be distinguished from those with general lung adenocarcinoma, and that in turn makes them candidates for a different chemotherapy regimen than the majority of patients with the malignancy.

EGFR is not only the name of the protein marker used to customize treatment but it is also the name of the gene that codes for the protein. The gene is mutated in about 10% of patients with nonsmall cell lung cancer and in about 50% of lung cancers that occur in patients who have never smoked [14]. These patients will usually benefit from a more personalized treatment regimen that includes erlotinib (Tarceva).

Memorial Sloan Kettering Cancer Center in New York uses gene sequencing to detect mutations in EGFR, as well as several other genetic mutations believed to contribute to a variety of rare and common cancers. Its genomic test panel, called MSK-IMPACT, has been used on patients with certain solid tumors, including melanoma and lung and colon cancer. It is now used on any solid tumor. This approach to oncology also rests on a relatively new theory that suggests the treatment of a specific cancer should not only be based on its location but also on its genetic makeup. The result has been that medication originally intended to combat breast or colon cancer, for example, may benefit patients with melanoma or a malignancy in some other organ for which the drug has not been approved because the seemingly unrelated malignancies share a similar genetic signature.

IS CANCER A GENETIC DISEASE?

There is abundant evidence to support the theory that cancer is a genetic disorder, but that does not necessarily mean all cancers have a genetic *cause*.

Tumor suppressor genes and oncogenes are involved in the pathogenesis of many malignancies. When tumor suppressor genes are performing their normal role, they are involved in what their name implies. When a cell's DNA is damaged, some of these genes are involved in repairing the genetic material. Others give instructions for cell death to eliminate the threat of damaged DNA disrupting the normal cell proliferation process. Although BRCA1 and BRCA2 are often discussed as contributing causes of breast cancer, both genes are actually tumor suppressor genes. It is only mutations in these genes that increase the risk of breast and ovarian cancer.

The role of oncogenes is also worth noting. In layperson's language: "Proto-oncogenes are genes that normally help cells grow. When a proto-oncogene mutates (changes) or there are too many copies of it, it becomes a 'bad' gene that can become permanently turned on or activated when it is not supposed to be. When this happens, the cell grows out of control, which can lead to cancer. This bad gene is called an oncogene" [15].

Despite this relationship between genes and cancer, a broader analysis of the human genomes calls into question the role of genes as the primary cause of the disease. Approximately 1000 genes have been associated with cancer, including about 250 oncogenes and 700 tumor suppressors [16]. David Wishart, from the departments of biological sciences and computing sciences at the University of Alberta, points out that a cell usually needs two or more mutations in a cancer-related gene to start the carcinogenic process. There may be more than 1 million unique cancer genotypes. Additionally, investigators have identified more than 2 million coding point mutations, more than 6 million noncoding mutations, about 61,000 genome rearrangements, and more than 60 million abnormal gene expression variants. Equally significant is the fact that whole genome sequencing has detected as many as 50,000 SNPs in tumor cells, when compared to normal tissue.

These statistics suggest that cancer may be *millions* of distinct diseases. And we cannot even be certain that all the detected genetic variants *drive* the cancer to develop. Many may simply be "along for the ride," so-called passenger mutations. The development of cancer may require between 2 and 20 driver mutations [17].

With all the variables in play, it may be better to look for a way to interrupt a few critical metabolic pathways that oncogenes and tumor suppressors give rise to. Many cancer-associated mutations have been found to affect three major metabolic pathways: aerobic glycolysis, glutaminolysis, and one-carbon metabolism [16]. This suggests that cancer may be a metabolic disease as much as a genetic disease. It also suggests that devising a therapeutic approach that addresses one or all of these metabolic aberrations would be more effective than addressing the millions of genetic defects. In recent years, the metabolic approach to cancer causation has resulted in the discovery of several metabolites associated with specific malignancies. The accumulation of these "oncometabolites" appears to initiate or sustain tumor growth and metastasis. The list includes 2-hydroxyglutarate, which is found in gliomas, a type of nervous system cancer, fumarate, which has been detected in renal cancer, sarcosine, detected in prostate cancer, glycine, located in breast cancer, and glucose, found in most cancers.

Pursuing a metabolic model also has implications for early detection and treatment. Wishart suggests that detecting levels of common metabolites in saliva, blood, and urine may serve as early signposts for various cancers, allowing earlier treatment. He also points out that some existing drugs that are known to target these metabolites may be repurposed as cancer agents:

Some existing drugs are already showing impressive results as anticancer therapies, including metformin (a diabetic biguanide that inhibits hexokinase II), dichloroacetate (a lactic acidosis drug that inhibits pyruvate dehydrogenase kinase), ritonavir (an antiviral drug that also inhibits glucose transporters) and orlistat (an anti-obesity drug that blocks fatty acid synthase). Likewise diets or medical foods that significantly reduce the amount of glucose (ketogenic diets) or the amount of non-essential amino acids have shown good promise in stopping or reducing tumor growth in animal models and even humans.

In many respects, cancer remains the proverbial elephant being examined by a group of blindfolded men and women trying to figure out what they are touching. One grabs the tail and concludes it is a genomic disorder, another tugs at its trunk and "sees" an infectious cause, while others examine the tusks, convinced it has a metabolic origin. But if cancer has a "holistic etiology," being caused by a long list of interacting contributing factors, a more sophisticated approach to its diagnosis and treatment will be required. A multifactorial etiology means each individual's cancer will be the result of mutations, metabolic dysfunction, psychosocial stress, nutritional inadequacies, exposure to environmental toxins, and a variety of other causes. And each of the causes likely contributes a different percentage to the genesis of the disease in each individual.

THE VALUE OF PHARMACOGENOMICS

Another contributing cause that will figure into the aforementioned holistic etiology is an individual's unique response to medication and food. Significant progress has been made on both fronts.

The list of Food and Drug Administration (FDA)-approved medications associated with a genetic variant is long. And the FDA has included warnings, precautions, or indication adjustments for each of these drugs [18]. The list takes into account germ line or somatic gene variants, functional deficiencies, gene expression changes, and chromosomal abnormalities. Many of these drugs have narrow therapeutic windows, which means that small changes in their absorption, metabolism, or excretion can have profound clinical consequences. And since many mutations affect the absorption, metabolism, or excretion of the drugs on the FDA list, ignoring the role of pharmacogenomics will put patients at risk of serious adverse effects or reduce the effectiveness of their medication.

Pharmacogenomic research has demonstrated that some individuals have a genetic variant responsible for the liver enzymes that metabolize warfarin (Coumadin), one of the most widely used anticoagulants. CYP4F2, one of the genes that are part of the cytochrome P450 family of genes, is involved in oxidation of vitamin K, which is the primary target of warfarin. Individuals who have inherited a variant of this enzyme produce a CYP4F2 enzyme that can slow down vitamin K oxidation, which in turn makes warfarin less effective as an anticoagulant, warranting larger doses [19]. Mutations in VKORC1, the gene responsible for an enzyme called vitamin k epoxide reductase subunit 1, have also been shown to alter a patient's response to warfarin. It is estimated that 30%–40% of the variations in response to warfarin can be attributed to genetic variants in these 2 genes.

These observations have been shown to have real-world clinical implications. In one study in which clinicians either took into account CYP2C9 and VKORC1 mutations or ignored these variants among patients being prescribed warfarin, hospitalizations for hemorrhage occurred 28% less often among the group whose their physicians paid attention to that information [19].

There is similar pharmacogenomics data on clopidogrel (Plavix). In fact, the package insert for the drug has a black box warning that states in bold print: DIMINISHED ANTIPLATELET EFFECT IN PATIENTS WITH TWO LOSS-OF-FUNCTION ALLELES OF THE CYP2C19. It goes on to say that the effectiveness of Plavix depends on conversion to an active metabolite by the cytochrome P450 (CYP) system, principally CYP2C19, and that tests are available to identify patients who are CYP2C19 poor metabolizers. Unfortunately, clinicians have been slow to apply such genetic data in everyday practice, in part because the Centers for Medicare and Medicaid Services (CMS) has been unwilling to provide reimbursement for many of the test procedures needed to confirm mutations in individual patients. Fortunately, CMS does cover CYP2C19 testing for clopidogrel, but the agency recently eliminated its coverage of many other genetically based drug sensitivity tests.

THE EMERGING SCIENCE OF NUTRITIONAL GENOMICS

There is also evidence to suggest a genetic component to nutrient requirements and food intolerances. Xiumei Hong and associates recently performed a GWAS that found regions on the HLA-DR and HLA-DQ genes on chromosome 6 (6p21.32) that are closely linked with allergies to peanuts [20].

A genetic basis for food intolerance also seems to exist among individuals and ethnic groups who have lactose intolerance. The inability to digest milk sugar exists among many populations, while the ability to fully digest the sugar is most common in Northwestern Europe, especially among Swedes and Danes. Lactase, the enzyme responsible for breaking down lactose in the intestinal tract, is encoded by the LCT gene, which is located on chromosome 2. Differences in individuals' ability to digest the sugar have been traced to a genetic polymorphism, which suggests that those who cannot convert lactose into glucose and galactose have inherited a recessive gene that causes lactase to decline after they are weaned[1] [21].

Genetics also plays a role in individuals' needs for micronutrients. A polymorphism in the gene responsible for the enzyme methylene tetra-hydrofolate reductase, for instance, has been found to severely disrupt the body's ability to metabolize the B vitamin folic acid. That in turn increases the threat of neural tube defects and cardiovascular disease. The mutation can be ameliorated with an increased intake of the vitamin.

Genetic variability apparently influences individuals' needs for vitamin C as well. Mutations that disrupt the synthesis of certain glutathione S-transferases increase a person's need for the vitamin, putting them at greater risk of a deficiency if they consume a diet that does not meet the Recommended Dietary Allowance for ascorbic acid [22].

There are also several inborn errors of metabolism that greatly increase the need for vitamin B_6 (pyridoxine) among those rare individuals with the gene variants.

- A mutation in chromosome 5, which affects an enzyme called α-aminoadipic semialdehyde dehydrogenase, causes pyridoxine-dependent epilepsy, requiring high doses of vitamin B6 to offset its effects.
- A mutation on chromosome 1 disrupts a second enzyme—tissue non-specific alkaline phosphatase—causing infantile hypophosphatasia, necessitating IV and oral vitamin B6 therapy
- A gene variant on chromosome 2 affects alanine/glyoxylate aminotransferase, causing primary hyperoxaluria Type I, likewise requiring vitamin B6 therapy.

To date, Philippa Mills and associates have identified 12 such genetic variants that increase the need for vitamin B_6 [23].

There have been contradictory studies suggesting that coffee intake increases the risk of myocardial infarction (MI). But when genetic variants among individuals are taken into account, the apparent contradictions are resolved. Caffeine is metabolized by a liver enzyme called cytochrome P450 1A2 (CYP1A2). A person who carries a mutated form of the gene responsible for the production of the enzyme—the mutated gene is called

---

[1] These individuals are homozygous for an autosomal recessive gene that causes lactose intolerance, which means they inherit the allele from both parents; the gene is autosomal, which means it is not transmitted on one of the sex chromosomes (X and Y), but by one of the 22 nonsex chromosomes. And the fact that it is recessive indicates that two copies, one from each parent, must be present for it to be expressed in the person's physiology. A dominant gene, on the other hand, only needs to be transmitted from one parent for it to be expressed.

CYP1A2★1F—slowly breaks down caffeine and is referred to as a slow metabolizer. A person who inherits two copies of the CYP1A2★1A gene, on the other hand, manufactures a form of the enzyme that rapidly metabolizes caffeine. (Inheriting two copies of a gene, one from each parent, is referred to as a homozygous carrier.) Marilyn Cornelis with the Department of Nutritional Sciences, University of Toronto, and colleagues evaluated the records of 2014 patients with acute MI and compared their intake and genetic makeup to that of 2014 matched controls. Their comparison revealed that 55% of MI patients and 54% of controls carried the slow metabolizer gene CYP1A2★1F. The more coffee the MI patients drank, the more likely they were to have had a heart attack. For example, the likelihood of a MI among patients drinking one cup a day was insignificant, when compared to patients who had drank four or more cups a day. (Odds ratio of a MI for 1 cup a day was 0.99; for 2–3 cups, it was 1.36; and for 4 or more cups, it was 1.64, which means drinking the most coffee increased the risk of an MI by about 64%.) On the other hand, among MI patients who quickly broke down caffeine, i.e., they were fast metabolizers who were homozygous for the CYP1A2★1A, drinking even large amounts of coffee had no effect on the incidence of an MI [24].

REFERENCES

[1] Feero WG, Guttmacher AE, Collins FS. Genomic medicine—an updated primer. N Engl J Med 2010;362:2001–11. http://www.nejm.org/doi/pdf/10.1056/NEJMra0907175.

[2] Framingham heart study. Genetic data. 2016. https://www.framinghamheartstudy.org/researchers/description-data/genetic-data.php.

[3] NIH National Human Genome Research Institute. Genome-wide association studies. August 27, 2015. https://www.genome.gov/20019523/.

[4] De Ligt J, Willemsen MH, van Bon BW, et al. Diagnostic exome sequencing in persons with severe intellectual disability. N Engl J Med 2012;367(20):1921–9. https://www.ncbi.nlm.nih.gov/pubmed/?term=10.1056%2FNEJMoa1206524.

[5] Partners Healthcare Personalized Medicine. Pan cardiomyopathy panel (62 genes) test details. http://personalizedmedicine.partners.org/Laboratory-For-Molecular-Medicine/Tests/Cardiomyopathy/PanCardiomyopathy-Panel.aspx.

[6] Empire Blue Cross Blue Shield. Genetic testing for diagnosis and management of hereditary cardiomyopathies. May 4, 2017. https://www.empireblue.com/medicalpolicies/policies/mp_pw_c132947.htm.

[7] Liao X, Lochhead P, Nishihara R, et al. Aspirin use, tumor PIK3CA mutation, and colorectal-cancer survival. N Engl J Med 2012;367(17):1596–606. https://www.ncbi.nlm.nih.gov/pubmed/?term=10.1056%2FNEJMoa1207756.

[8] Saunders CJ, Miller NA, Soden SE, et al. Rapid whole-genome sequencing for genetic disease diagnosis in neonatal intensive care units. Sci Transl Med 2015;4(154):154ra135. https://www.ncbi.nlm.nih.gov/pmc/articles/PMC4283791/.

[9] DeVlaminck I, Valantine HA, Snyder TM, et al. Circulating cell-free DNA enables non-invasive diagnosis of heart transplant rejection. Sci Transl Med 2014;6(241):241ra77. https://www.ncbi.nlm.nih.gov/pubmed/24944192.

[10] Rebbeck TR, Mitra N, Wan F, et al. Association of type and location of BRCA1 and BRCA2 mutations with risk of breast and ovarian cancer. JAMA 2015;313(13):1347–61.

[11] Maas P, Barrdahl M, Joshi AD, et al. Breast cancer risk from modifiable and nonmodifiable risk factors among white women in the United States. JAMA Oncol 2016. May 26, Epub ahead of print. https://www.ncbi.nlm.nih.gov/pubmed/27228256.

[12] Khoury MJ. Using genomics in precision prevention of breast cancer. CDC Genomics and Health Impact Blog; April 23, 2015. https://blogs.cdc.gov/genomics/2015/04/23/using-genomics/.

[13] Cardoso F, van't Veer LJ, Bogaerts J, et al. 70-Gene signature as an aid to treatment decisions in early-stage breast cancer. N Engl J Med 2016;375(8):717–29.

[14] Memorial Sloan Kettering Cancer Center. Lung cancer/lung cancer diagnosis: genomic testing. Accessed July 17, 2017. https://www.mskcc.org/cancer-care/types/lung/diagnosis/genetic-testing.

[15] American Cancer Society. Genes and cancer: oncogenes and tumor suppressor genes. June 24, 2014. http://www.cancer.org/cancer/cancercauses/geneticsandcancer/genesandcancer/genes-and-cancer-oncogenes-tumor-suppressor-genes.

[16] Wishart DS. Is cancer a genetic disease or a metabolic disease? EBioMedicine 2015;2:478–9.

[17] Wapner J. Cancer gene tests provide few answers. Sci Am September 2016:24–6.

[18] Food and Drug Administration. Table of pharmacogenomic biomarkers in drug labeling. July 11, 2016. http://www.fda.gov/Drugs/ScienceResearch/ResearchAreas/Pharmacogenetics/ucm083378.htm.

[19] Wang L, McLeod HL, Weinshilboum RM. Genomics and drug response. N Engl J Med 2011;364:1145–53.

[20] Hong X, Hao K, Ladd-Acosta C, et al. Genome-wide association study identifies peanut allergy-specific loci and evidence of epigenetic mediation in US children. Nat Commun 2015;6:6304. https://www-ncbi-nlm-nih-gov.ezp-prod1.hul.harvard.edu/pubmed/?term=Genome-wide+association+study+identifies+peanut+allergy-specific+loci+and+evidence+of+epigenetic+mediation+in+US+children.

[21] Swallow DM. Genetics of lactase persistence and lactose intolerance. Annu Rev Genet 2003;37:197–219.

[22] Cahill LE, Fontaine-Bisson B, El-Sohemy A. Functional genetic variants of glutathione S-transferase protect against serum ascorbic acid deficiency. Am J Clin Nutr 2009;90:1411–7.

[23] Mills PB, Footitt EM, Clayton PT. Vitamin B6 metabolism and inborn errors. In: Valle D, editor in chief. The online metabolic and molecular bases of inherited disease. New York: McGraw Hill; 2016. http://ommbid.mhmedical.com/book.aspx?bookID=971.

[24] Cornelis MC, El-Sohemy A, Kabagambe EK, et al. Coffee, CYP1A2 genotype, and risk of myocardial infarction. JAMA 2006;295:1135–41.

CHAPTER FOUR

Small Data, Big Data, and Data Analytics

Data analytics is as old as medicine itself.

When British physician John Snow was trying to determine the cause of the deadly cholera epidemic in London, he mapped the occurrence of the disease throughout the streets of London. One of his data collections, called "Showing the Mortality from Cholera, and the Water Supply, in the Districts of London, in 1849," provided a detailed tabulation of the outbreaks, deaths, locations, and water suppliers, enabling him to ascertain that wells contaminated with fecal matter were causing the epidemic [1].

When Ignaz Semmelweis, a Hungarian physician, was trying to determine why obstetric patients examined by medical students were dying of childbed fever, while women delivered by midwives were not, he too collected the relevant data; in this case, mortality rates in the first and second clinics at the Vienna General Hospital. His "small data" suggested that the greater number of deaths among women examined by medical students occurred because they had been dissecting cadavers before they examined their patients, without disinfecting their hands before examining the women.

What made these landmark discoveries so outstanding was not the pen and paper used to collect the relevant statistics, but the ability of two bold, independent-minded clinicians to see *patterns* where others saw noise.

Now that we have entered the world of Big Data, the landscape has not changed as much as many enthusiasts would like to imagine. Pen and paper have been replaced with impressive tools such as parallel computing, data warehouses, and sophisticated statistical methods. But what has not changed is the need for bold, independent-minded clinicians and data scientists to recognize patterns and insights that others fail to see.

Our goal in this chapter is to review the 21st century analytical tools now at our disposal, which obviously give us an advantage that our 19th century colleagues did not have. But equally important, we focus much of our attention on published research that discusses not just the "potential" for benefit but actionable insights that will improve patient care.

Realizing the Promise of Precision Medicine
ISBN 978-0-12-811635-7
http://dx.doi.org/10.1016/B978-0-12-811635-7.00004-X

UNDERSTANDING THE LANGUAGE OF DATA ANALYTICS

An understanding of how data analytics impacts clinical medicine requires at least a basic understanding of the terminology and infrastructure of Big Data. Big Data is usually distinguished from "small data" by its volume, velocity, and variety. The so-called 3-Vs reflect the fact that the amount of data available for analysis is huge, compared to the quantity of data that has traditionally been used in clinical trials and epidemiologic studies. The databases currently being examined may include petabytes of data, each of which contains 1024 terabytes, or exabytes, which consist of 1024 petabytes each. (A terabyte contains 1024 gigabytes.) They may include billions of patient records, including electronic health record (EHR) data, social media, claims data, and much more. They typically consist of structured data—for example, the ICD (International Classification of Diseases) codes for patients—and unstructured data such as narratives describing patients' signs and symptoms. The speed or velocity with which these data are accumulating and at which they can be moved from place to place also distinguishes Big Data from more traditional source of patient information, as does the variety of types of data, which can include input from remote sensors, text data on hard drives and smartphones, imaging data from videos, photographs, and X-rays, and so on.

All the data can be stored in different types of "storage bins," including relational databases and data warehouses, and analyzed by linking these storage bins through a process called distributed computing, using a tool such as Hadoop. Data scientists also use terms such as semantics, syntax, and ontology, all of which have specialized meanings in the context of medical informatics.

In a simple computer file folder or directory, you might store individual files with information as Microsoft Office text documents, pdfs, and images as tiff or jpeg files. But if one were to try to determine relationships, correlations, or patterns between these diverse collections of data, it would be difficult. Relational databases help solve this problem, making analysis easier by creating a schema, which is the structural representation of the data in the database. The data are compartmentalized in various tables, fields, rows, and columns, and the database creates relationships among the numerous data points. One can then query the database to pull out links between columns, rows, etc. One table may list all the demographic details on patients, including age, address, gender, and race, and a second may list their family medical history. Querying the database, you may be able to determine which female patients have a history of a specific disease because the database already has links built into it between the demographics table and the family history table.

Analyzing the structured data in a relational database has merit; but unfortunately, a great deal of the data collected during clinical medicine is unstructured and therefore cannot be analyzed the same way. In recent years, a variety of other digital tools have been created to render these data more analysis friendly, including natural language processing (NLP). NLP uses algorithms to extract useful data from a narrative account, for instance, and then puts it into a structured setting so that it can be analyzed. Obviously this short primer on relational databases and NLP only scratches the surface and vastly oversimplifies data analytics.

Understanding the concept of distributed computing is another piece of the puzzle that can make Big Data less mysterious. In simplest terms, distributed computing is a way for individual computers to talk to one another and function as one gigantic "brain" despite the fact that these computers may be located all over the world. The Internet is a distributed computing network, connected by nodes, routers, and the like. Hadoop (pronounced hə'duːp/) is another. It is currently being used by data analysts to gain insights from massive amounts of data that are too large to be stored economically in any one location. *Big Data for Dummies* offers a succinct definition: "An Apache-managed software framework … [that] allows applications based on MapReduce to run on large clusters of commodity hardware. Hadoop is designed to parallelize data process across computing nodes to speed computations and hide latency" [2]. Of course, a definition that itself includes somewhat mysterious terms is problematic as well.

Apache refers to software used in web servers. Parallel processing is a way of handling data by splitting it into several parts, with each part being processed at the same time by different microprocessors. In a single computer, the data analysis would be processed by two or more chips within the same machine. In a Hadoop-driven data analysis, the advantage is that the analyst now has numerous high-powered microprocessors scattered throughout the entire computer network. MapReplace is a software framework that allows analysts to execute a series of functions on a large collection of data spread across all the machines in the network. The *Mapping* aspect of this tool refers to its ability to distribute the problem to be solved across a large number of computer systems, balancing the workload among machines in the network so that no one machine is overloaded. Once an analytic function has been performed on the data, the *Reducing* aspect of the framework aggregates the results, putting things back together again into a readable whole. More details on MapReplace can be found on IBM's website.

The other set of concepts that are useful to understand include semantics, syntax, and ontology as they apply to computers. None of the terms make much sense unless you remember that like humans, computers have their own *languages*, and like any human language, computer languages have to have structure. In the English language, syntax includes the set of rules that determine the order of the words in a sentence. Listening to the way the Jedi master Yoda constructs his sentences makes the point. When Luke Skywalker says he is looking for someone, Yoda replies: "Found someone, you have, I would say, hmmm?", ignoring the standard position of words in English.

Similarly, computer languages require the characters be positioned a certain way in a line of code. This syntax is a set of rules that defines the proper combination of letters and other symbols that a computer will respond to in the correct way. The syntax for email addresses, for example, requires the @ symbol and the dot (.) to be inserted at specific locations in the address to send a message. Anyone who has had the experience of trying to send an email that left out the @ symbol realizes the need for proper syntax.

In a semantic error, on the other hand, you may enter a legal command to the computer, using the right syntax, but its *meaning* makes no sense in the program you are using. Semantics as it applies to computing programming is beyond our scope. In medical informatics, however, semantics is an important issue. It is especially important to agree on the meaning of the terms we are using, and how they are categorized in various structured hierarchies. That brings up the term ontology and the computer languages that are essential for digital medicine to work.

In its simplest sense, ontology is about putting things into categories. And in medicine, those categories have to be precise and agreed upon by everyone to communicate patient data correctly between providers. Several standardized vocabularies have been created to help classify and categorize the language of medicine and to help clinicians and administrators communicate the complexities of medicine via computers. Among the more important ones are SNOMED-CT, LOINC, and RxNorm. SNOMED-CT, which stands for Systematized Nomenclature of Medicine—Clinical Terms, standardizes the clinical terms used in EHRs and elsewhere so that patient data can be exchanged accurately. LOINC, or Logical Observation Identifiers Names and Codes, fulfills a similar role in communicating laboratory test results and measurements, covering areas such as chemistry, hematology, and microbiology. RxNorm spells out the terminology needed to communicate about medications in the US market.

ALGORITHMS AND MACHINE LEARNING

No discussion about data analytics would be complete without discussing algorithms. Fundamentally the term refers to a set of rules for solving a problem, but as clinicians and technologists know, these rule sets can be ridiculously simple or painfully complex. Physicians frequently use decision tree algorithms to arrive at a diagnosis or to decide on an evidence-based treatment plan. Similarly, clinical decision support systems (CDSS) may rely on computer-based algorithms that are derived from nationally accepted practice guidelines, metaanalyses, systematic reviews, randomized clinical trials (RCTs), and expert opinions to help guide patient care. These tools take existing rules and apply them to individual circumstances. But a new breed of decision-making aids are now emerging that make use of more advanced algorithms that teach computers to learn, to adapt to massive amounts of raw data that it interprets, and then to use these revelations to generate new rules.

Welcome to the world of machine learning. In traditional expert-based systems, the program typically contains general principles, which the clinician then applies to specific situations. In machine learning, the computer uses sophisticated algorithms to analyze thousands or millions of data points to *create* rules, moving from the specific to the general. Once the new tool has learned from all the input, it can then be used to address specific situations.

Machine learning has made significant advances in radiology. No longer are we limited to basic interpretations of X-rays such as atelectasis or effusion. With enough computing power, it is possible for the right algorithms to analyze the millions of pixels beneath the radiograph, looking for subtle patterns and detecting parenchymal opacities and other anomalies missed by the naked eye [3]. In fact, Ziad Obermeyer, MD, from Harvard Medical School, and Ezekiel Emanuel, MD, PhD, from the University of Pennsylvania believe [3] the following:

Machine learning will displace much of the work of radiologists and anatomical pathologists. These physicians focus largely on interpreting digitized images, which can easily be fed directly to algorithms instead. Massive imaging data sets, combined with recent advances in computer vision, will drive rapid improvements in performance, and machine accuracy will soon exceed that of humans. Indeed, radiology is already partway there: algorithms can replace a second radiologist reading mammograms and will soon exceed human accuracy.

A recent paper published in the *Journal of the American Medical Association* also suggests that some of the diagnostic burden that now rests on the shoulders of ophthalmologists may be relieved by machine learning as well. Varun Gulshan, PhD, with Google, Inc., and associates employed a type of machine learning called deep learning, which uses back propagation, a type of optimization algorithm, to teach a computer to recognize diabetic retinopathy by analyzing more than 128,000 retinal fundus photographs [4]. Gulshan et al. describe deep learning as "the process of training a neural network (a large mathematical function with millions of parameters) to perform a given task. The function computes diabetic retinopathy severity from the intensities of the pixels in a fundus image." The images had been previously evaluated by a panel of experienced ophthalmologists and residents as a means of comparison.

During the development or training stage of the study, the computer program identified 33,246 cases of referable diabetic retinopathy from a total collection of assessable images of 118,419 (28.1%). During the validation stage of the investigation, the software was tasked with interpreting photos from 2 data sets: EyePACS-1, which included 9963 images, and Messidor 2, which included 1748 images. The final analysis revealed that the program could detect the disorder with a sensitivity rating between 87% and 97.5% and a specificity of 93.4%–98.1%. Although these results are impressive, Gulshan and colleagues caution that the program still needs to be evaluated in a clinical setting to prove that it actually improves outcomes when compared to ophthalmologic assessment by clinicians.

PUTTING DATA ANALYTICS TO THE TEST

One on the most important Big Data projects to have a clinical impact was conducted by David J. Graham, MD, MPH, the Food and Drug Administration's (FDA's) Associate Director for Science, Office of Drug Safety, and his colleagues. Their investigation analyzed the patient records of approximately 1.4 million patients who belonged to Kaiser Permanente in California. Their aim was to determine if rofecoxib (Vioxx) increased the risk of acute myocardial infarction (MI) and sudden cardiac death. Unlike conventional case-control studies, Graham et al. were able to review more than a million patient records, the equivalent of 2,302,029 person years of follow-up. In this population, they detected 8142 cases of serious coronary heart disease (CHD), 2210 of which were fatal. The odds of developing CHD among patients taking any dose of the medication were 59% greater than it was among controls (95% CI 1.10–2.32, $P = .01$) [5]. In patients who took 25 mg/day or less, the odds were 47% greater for developing CHD (95% CI 0.99–2.17, $P = .054$). Finally, among patients who took high doses, namely more than 25 mg daily, the odds of heart disease were 258% greater (95% CI 1.27–10.11, $P = .016$).

Before their data were presented, several earlier smaller studies suggested an association between this cyclo-oxygenase 2 (COX-2) selective nonsteroidal antiinflammatory drug and heart disease, but the findings have several shortcomings. The data from Graham et al., which had been presented at a conference before being published in *The Lancet,* made headline news and embroiled the researchers in a confrontation with FDA officials, who initially did not want the results made public [6]. The drug was withdrawn from the market by Merck on September 30, 2004. But it has been estimated that during the time the drug was on the market, more than 100 million prescriptions were written and between 88,000 and 140,000 excess cases of serious CHD may have resulted from the public's exposure. Its removal was clearly a testament to the value of data analytics.

The Graham et al. study is only one of many that illustrate the benefits of using massive data sets to detect meaningful insights in health care. Their study looked back in time, comparing cases of heart disease to controls. But data analytics has also been helpful in the emerging field of predictive analytics, looking forward in time to pinpoint the existence of risk factors that predispose patients to a variety of disorders or adverse outcomes. In Chapter 1, for example, we discussed an analysis performed by Jeremy Sussman and his colleagues at the University of Michigan and Tufts Medical Center.

Sussman et al. [7] analyzed data from more than 3000 patients who had participated in the Diabetes Prevention Program (DPP), which had divided patients into three groups: A placebo group that received standard lifestyle advice, an intensive lifestyle modification group, and a group that received an oral antidiabetic drug (metformin) plus standard lifestyle advice. The DPP study found that the intensive lifestyle group saw the most benefit, with the lowest incidence of diabetes over time, while the medication group came in second place. But Sussman et al. were interested in determining how these results would apply to *individuals* who were at risk for diabetes.

The researchers found "that average reported benefit for metformin was distributed very unevenly across the study population, with the quarter of patients at the highest risk for developing diabetes receiving a dramatic benefit (21.5% absolute reduction in diabetes over three years of treatment) but the remainder of the study population receiving modest or no benefit." By way of contrast, the difference in benefit from the intensive lifestyle training between higher and lower risk patients was minimal.

With these results in mind, Sussman and his associates developed a diabetes risk prediction tool to help identify which individuals may be most likely to develop hyperglycemia, and which patients would experience little or no benefit but would be exposed to the possible adverse effects of treatment. That analysis generated a risk prediction tool that may help clinicians apply the data from the DPP to individual patients. (More details on that tool are available on page 12, Fig. 1.2). The reanalysis of the DPP was based on an approach to personalized medicine sometimes referred to as benefit-based tailored treatment [8].

If personalized medicine is to have a lasting impact on clinical medicine, such reanalysis of large RCTs will need to be performed more frequently. It has been estimated, for example, that among patients at risk for acute MI, the mortality rate can vary as much at 10-fold, when patients in the highest baseline risk quartile are compared to those in the lower risk quartile. In clinical trials involving patients who are HIV positive, that disparity in risk among highest and lowest quartiles can be as great as 46-fold [8,9].

Traditionally, investigators have performed subgroup analyses to detect smaller subpopulations within a study population who deviate from the group as a whole and for whom the study's conclusions may not apply. But subgroup analyses often fall short for two reasons: The number of patients in a subgroup may be too small to generate statistically meaningful results. And second, the analysis usually only explores one variable at a time. For instance, if a study concludes that statin A reduces low-density lipoprotein cholesterol by 25% on average for the whole population, researchers may pull out the data on patients older than 75 to determine if the treatment effect remains as robust. But the differences in the treatment effect may only result from several intertwined risk factors, with age being only one of them, thus the need for a multivariate analysis. In a scenario like this, pulling age from the picture would be like removing a single thread from an intricate, multicolored tapestry: The entire tapestry may easily come apart.

Typically a multivariate risk analysis that stratifies patients is sufficiently statistically powered to detect treatment differences and to establish that the study population is heterogeneous. In other words, it can bring us closer to personalizing RCT results.

As the results of Sussman et al.'s research indicate, this is not just an academic exercise in statistics. Similarly several other multivariate risk-stratified analyses have demonstrated that summary results from large clinical trials need to be reanalyzed for them to be useful to clinicians in community practice. Such analyses can have an impact on how we evaluate the benefits of carotid endarterectomy, tirofiban in non-ST-elevation acute coronary syndromes, primary angioplasty, and other modalities [10–12].

Unfortunately, performing the type of data analytics needed to make clinical research more personalized is still not a priority for most investigators. As David M. Kent, MD, and Rodney A. Hayward, MD, point out in their critique: "Multivariate risk-stratified analyses based on easily obtainable clinical variables are often informative and frequently feasible, but rarely performed. Making such analyses standard should be seriously considered" [8].

Like endocrinologists who manage prediabetic patients, pediatricians also must cope with at-risk populations and the uncertainties about over- and undertreatment. The decision to treat newborns at risk of early onset sepsis is one such conundrum. Current Centers for Disease Control and Prevention (CDC) guidelines are somewhat vague on how to identify infants at high enough risk to justify empirical antibiotic therapy. One analysis found that 13% of both well-appearing and ill-appearing infants were evaluated for early onset sepsis. More than 1 out of 10 were treated with antibiotics before the sepsis could be confirmed. But once blood cultures arrived, sepsis was confirmed in only 0.04% of the group [13]. With so many infants receiving unwarranted medication, Escobar et al. decided to develop a stratification algorithm to help identify those at high and low risk for sepsis. Starting with more than 600,000 infants (≥34 weeks gestation) at 14 hospitals between 1993 and 2007, they identified 350 cases of early onset sepsis; 1063 infants served as controls. Using recursive partitioning and logistic regression, they were able to categorize infants into three risk groups based on maternal risk factors and a newborn's changing clinical status, thereby helping clinicians to classify infants into either a (1) treat empirically, (2) observe and evaluate, or (3) continue observation group. The analysis suggested that only 4.1% of the infants in their cohort would require antibiotics, while 11.1% could be observed and evaluated. 84.8% fell into the continue observation group. Their conclusion: "Judicious application of our scheme could result in decreased antibiotic treatment in 80 000 to 240 000 US newborns each year."

Data analytics is also proving useful in helping to manage hospital readmissions. Section 3025 of the Affordable Care Act established the Hospital Readmissions Reduction Program, which reduces reimbursements to hospitals that allow excessive, preventable 30-day readmissions. The initial program was applied to readmission for acute MI, heart failure (HF), and pneumonia [14]. (The Agency subsequently added total hip and knee replacements and chronic obstructive pulmonary disease to its list of preventive causes of readmissions.) Since that program went into effect, which began with hospital discharges on October 1, 2012, hospitals have been struggling to develop ways to reduce their avoidable readmissions.

A program implemented at Parkland Health and Hospitals System in Dallas has seen success in the effort. It employed a suite of computerized case monitoring and case coordination programs to address the readmissions issue. The electronic medical records-based platform stratified patients who were admitted with HF, categorizing them with the help of an electronic predictive model that estimated their risk of being readmitted within 30 days. The predictive tool was able to reduce readmissions from 26.2% to 21.2%, a significant improvement that persisted after adjustment for confounding variables, with an adjusted odds ratio of 0.73 [15]. In practical terms, that meant 45 fewer readmissions occurred among 913 patients enrolled in the experiment.

The predictive tool used to arrive at these results relied on 29 clinical, social, behavioral, and utilization factors that were extracted from the electronic medical record (EMR) within 24 h of each patient's admission for HF. Once the patient's EMR data was extracted and he or she was identified as having HF, the software notified the appropriate personnel, who then applied a series of inpatient and outpatient protocols to address the specialized needs of this patient population. They included HF-relevant counseling and monitoring based on readmission reduction strategies that have been well-documented in the medical literature. The interventions included a follow-up phone call by a nurse within 48 h of discharge, individualized case management services for 30 days, and an appointment with a cardiologist within 7 days of discharge. Obviously the success of this program required considerable clinical and administrative resources, but without the data analytics and the software, many of the high-risk patients would likely have fallen through the cracks. Amarasingham et al. sum up their results succinctly: "We found that a care transition intervention that directed largely existing resources to a smaller subgroup of patients with HF based on daily EMR-based risk stratification produced a clinically meaningful reduction in overall readmissions. By concentrating care management efforts on about one-quarter of patients with HF we were able to demonstrate a 26% relative reduction in the odds of readmission and an absolute reduction of 5.0 readmissions per 100 index admissions." The data analytics initiative at Parkland Health and Hospitals System is only one of many projects that is generating actionable insights on how to manage patients who are at risk of avoidable hospital readmission.

Once again we want to emphasize the ability of data analytics to identify *subgroups* of at-risk patients. While personalized medicine may at times pinpoint the needs of *individual* patients, it often succeeds by narrowing down the number of patients requiring specialized services by identifying smaller segments of a population that would ordinarily be overlooked. Equally important is the ability of such analysis to eliminate portions of a patient population that do not require specialized services, which in turn reduces health-care costs and helps prevent medical staffs from being overutilized.

THE ROLE OF BIG DATA IN HYPOTHESIS TESTING

Analyzing massive volumes of health-care data can also help researchers avoid errors that sometimes challenge researchers who are looking for statistical significance. A Type I or alpha error occurs when a study concludes that there is a significant difference between two groups—a control group and an experimental group on a new drug, for example—when no difference actually exists. A Type II or beta error occurs when a study concludes that there is no real difference between treatment and control groups when in fact a true treatment effect exists. One reason a Type II error occurs is because too few subjects were included in the study.

Over the last several decades, there have been numerous examples of false negative studies that concluded that a specific treatment protocol was useless when in fact that conclusion was unwarranted. These Type II errors hinder innovation. Freiman et al. documented the publication of 71 "negative" randomized trials that arrived at that unjustified conclusion. Frieman and associates found that the sample size in these studies was not large enough to give a high probability (>90%) of detecting a 25% and 50% therapeutic improvement. The investigators concluded that "Many of the therapies labeled as 'no different from control' in trials using inadequate samples have not received a fair test" [16]. A second analysis of the research literature, published 14 years later, found that the same mistake was still quite common. Moher et al. reviewed 383 RCTs and found that most of the studies with negative results did not have large enough sample sizes to detect a 25% or 50% difference between experimental and control groups [17].

Tapping into the massive databases now available to many researchers can help reduce the likelihood of these false negative studies by increasing sample sizes the investigators have access to.

For example, there is some evidence to suggest that the potassium-sparing diuretic triamterene also has a blood pressure (BP) lowering effect.

But several studies have failed to demonstrate that it significantly lowers BP, despite the fact that there is a plausible mechanism of action for believing the drug should have such an effect. Wanzhu Tu, PhD, from the department of biostatistics at the Indiana School of Medicine, and colleagues gathered data on more than 17,000 patients with hypertension and compared the BP of those who were taking only hydrochlorothiazide, a diuretic with proven antihypertensive effects, to those taking hydrochlorothiazide plus triamterene. The 17,000+ patient records were derived from the Indiana Network of Patient Care, an electronic clinical information exchange. They found that patients on the drug combination had BP readings that were 3.8 mmHg lower than those on hydrochlorothiazide alone. Previous smaller studies, including a Cochrane Review of 6 studies, which included 2 trials with fewer than 150 patients each, had been unable to detect this effect [18].

The field of epidemiology is also slowly embracing the analysis of large data sets to help supplement more traditional approaches to infectious disease surveillance. Currently, epidemiologists and public health officials track the development of infectious diseases using laboratory-based surveillance and notifiable disease surveillance. Lab-based surveillance relies on data from lab results from hospital and outpatient facilities that confirm the presence of pathogens in patient specimens—blood, urine, and so on. The National Notifiable Diseases Surveillance System, part of the CDC, is a nationwide collaboration that allows local, state, federal, and international authorities to share data on diseases that clinicians are required to report to authorities when they see patients with these conditions.

In the 19th century, physicians like John Snow were still using pen and paper to report cases of infectious diseases. The 20th century has given us computerized reporting systems to improve the process, but the specialty has yet to take full advantage of the emergence of EHRs, Internet search queries, and social media. Many experts have been cautious about embracing these sources since the unsuccessful introduction of Google Flu Trends (GFT). GFT made headlines when it first surfaced, but its attempts to track influenza outbreaks using a specially designed algorithm to analyze Internet search queries failed to provide reliable data. GFT missed the initial influenza A (H1N1) pandemic in New York City in 2009, for example, and overestimated the severity of the 2012–13 flu epidemic [19].

Using the medical claims data that are generated with patient visits may be more promising. For example, electronic medical claims data compiled by IMS Health from 480 US locations have been used to monitor influenza-like illness between 2003 and 2010. These data, which represented 62% of outpatient visits in 2009, were compared to surveillance data sets from the CDC's-confirmed outpatient and lab-confirmed flu data, yielding a Pearson correlation coefficient of ≥ 0.89 [20] (see Fig. 4.1). The medical claims data accurately captured weekly changes in flu activity in the United States between and during pandemic seasons, including the first pandemic in 2009.

Even more promising is the development of hybrid tracking systems that combine traditional methods of collecting surveillance data such as lab-confirmed cases of infection with new approaches, which can include Internet search queries, Twitter feeds, and medical claims data. The CDC, for instance, is now collecting flu activity data that go beyond simple monitoring in the hope that it can forecast influenza outbreaks. The Agency launched its Flu Activity Forecasting site on January 19, 2016, and has enlisted the help of academic researchers who are using new digital tools to help make flu forecasting a viable supplement to existing surveillance methods [21].

Although the aforementioned studies cast Big Data in a positive light, blind faith in the power of data analytics is unjustified. As Voltaire once said: "Doubt is not a pleasant condition, but certainty is an absurd one." The same can be said about Big Data. A case in point: The Veterans Health Administration has performed advanced data analytics over the years to improve the quality of care given US veterans, and to meet the performance metrics being increasingly demanded of health-care organizations. The agency created a Corporate Data Warehouse to house patient data in 2006 and has developed risk scores to help predict which veterans are most likely to require hospitalization. One review of their operations stated the following: "Accessed 3000 to 4000 times monthly by more than 1200 clinicians, these scores are widely used in practice. Nurse care managers used these scores to guide services, including end-of-life and palliative care, delivered by multidisciplinary patient-aligned care teams (PACTs) to high-risk individuals. Compared with 87 practices with the lowest implementation of PACTs, the 77 practices with highest PACT implementation demonstrated a 17% reduction in hospitalizations (4.42 vs. 3.68 quarterly admissions per 1000 veterans) for ambulatory care–sensitive conditions and a 27% reduction in emergency department visits (188 vs. 245 visits per 1000 patients) over a 7-month period" [22].

But a review of the Veterans Administration's data analytics initiatives by Stephen Fihn, director of the Veterans Health Administration Office of Analytics and Business Intelligence, and his associates also acknowledged

Correlation outcome	Region 1 (Boston)	Region 2 (New York City)	Region 3 (Wash. DC)	Region 4 (Atlanta)	Region 5 (Chicago)	Region 6 (Dallas)	Region 7 (Kansas City)	Region 8 (Denver)	Region 9 (San Francis.)	Region 10 (Seattle)	Avg. (95% CI)
IMS-ILI v. CDC-ILI											
Wkly inc. (lag)[a]	**0.94 (1)**	**0.88 (0)**	**0.90 (1)**	**0.93 (1)**	**0.97 (1)**	**0.90 (1)**	**0.94 (1)**	**0.90 (1)**	**0.90 (1)**	**0.83 (0)**	**0.91 (0.88; 0.93)**
Peak week	**0.93**	**0.95**	**0.87**	**0.94**	**0.98**	**0.95**	**0.99**	**0.70**	**0.99**	**0.80**	**0.91 (0.85; 0.97)**
Seas. intensity	**0.92**	**0.63**	0.24	**0.62**	**0.62**	0.09	**0.76**	0.19	0.54	0.54	**0.52 (0.33; 0.68)**
IMS-ILI v. CDC viral surveillance											
Wkly inc. (lag)[b]	**0.89 (1)**	**0.87 (1)**	**0.89 (1)**	**0.93 (1)**	**0.85 (0)**	**0.90 (1)**	**0.88 (0)**	**0.93 (1)**	**0.80 (1)**	**0.94 (1)**	**0.89 (0.86; 0.91)**
Peak week	**0.93**	**0.97**	**0.97**	**0.92**	**0.94**	**0.99**	**0.99**	**0.98**	**0.99**	**0.97**	**0.97 (0.95; 0.99)**
Seas. intensity	0.06	0.15	0.05	-0.02	-0.27	0.66	0.23	0.34	-0.18	0.10	0.11 (-0.05; 0.28)
CDC-ILI v. CDC viral surveillance											
Wkly inc. (lag)[c]	**0.91 (0)**	**0.83 (0)**	**0.83 (0)**	**0.92 (0)**	**0.85 (0)**	**0.86 (0)**	**0.89 (0)**	**0.90 (0)**	**0.71 (1)**	**0.82 (1)**	**0.85 (0.81; 0.89)**
Peak week	**0.98**	**0.86**	**0.94**	**0.96**	**0.96**	**0.95**	**0.97**	**0.76**	**0.99**	**0.79**	**0.92 (0.87; 0.97)**
Seas. intensity	0.35	0.54	0.62	0.58	-0.02	0.33	0.75	0.81	0.38	0.52	0.49 (0.34; 0.63)

Values indicate Pearson correlation coefficients; values in bold are significant. For ILI time series, seasonal intensity is based on excess incidence over baseline each season, estimated from Serfling seasonal regression. For CDC laboratory surveillance, seasonal intensity is based on the cumulative percent virus positive each season (sum of influenza virus positives/sum of respiratory specimens tested).
[a] Lag maximizing the correlation between the two indicators indicated in parentheses. A positive lag indicates that IMS-ILI surveillance is ahead of CDC-ILI.
[b] A positive lag indicates that IMS-ILI is ahead of CDC viral surveillance.
[c] A positive lag indicates that CDC-ILI surveillance is ahead of CDC laboratory-confirmed viral activity.
doi:10.1371/journal.pone.0102429.t001

Figure 4.1 IMS Health from US locations was used to monitor influenza-like illness between 2003 and 2010. These data were compared to surveillance data sets from the CDC's-confirmed outpatient and lab-confirmed flu data. Medical claims data accurately captured weekly changes in flu activity in the United States between and during pandemic seasons, including the first pandemic in 2009 [20].

that the agency has fallen short. Its overemphasis of performance metrics has served as "the incentive to distort or game reporting of data. A recent, unfortunate manifestation of this phenomenon has been the recognition that a complex set of metrics created within the VHA to ensure that patients had prompt access to care were instead being misapplied or artfully manipulated to conceal an underlying lack of capacity to meet arbitrary performance targets" [23]. And with the revelation in recent years that thousands of veterans may have experienced life-threatening delays in receiving patient care, there is no doubt that data analytics is almost worthless if not accompanied by accountability and transparency on the part of those who manage a health-care organization.

RECONCILING BIG DATA AND RANDOMIZED CONTROLLED TRIALS

The debate about the value of randomized controlled trials has been going on for decades. Although they are considered the gold standard for verifying the validity of a contested treatment option, RCTs have significant limitations. They are very expensive to perform; the results sometimes do not apply to individual patients in community practice because the inclusion and exclusion criteria that existed when the RCT was conducted do not reflect real-world conditions; and some treatments cannot be adequately tested in RCTs for ethical reasons or because there are too few subjects available, which is usually the case with rare diseases or uncommon adverse drug reactions.

Unfortunately, many thought leaders in clinical medicine believe the only justification for recommending a treatment protocol is the RCT. They insist that since the RCT is the most effective way to arrive at scientific truth, any less rigorous evidence cannot be used as the basis for recommending therapy. This black or white approach to reliable medical treatment deprives patients of many options that have persuasive but not definitive evidence to support them. Among these "silver-standards" and "bronze-standards" are open clinical trials that do not include a placebo arm, case-control studies, animal experiments, test tube studies that confirm the mechanism of action for a therapeutic agent, and observational data derived from Big Data analytics.

The Sussman study discussed earlier in this chapter illustrates the shortcomings of the RCT, in this case the DPP study, which assigned 3234 patients with elevated plasma glucose levels to placebo, metformin, or a lifestyle modification program. Sussman et al. developed a predictive analytics scheme to make the results of the DPP study practical and personalized. Similarly, the Big Data-enabled Graham study discussed above was required to firmly establish the relationship between rofecoxib and heart disease because earlier studies, including the Vioxx Gastrointestinal Outcomes Research or VIGOR study, also a RCT, were unable to establish the link [24].

The Sussman and Graham studies are only two of the Big Data-enabled investigations that are now supplementing the findings of RCTs, and in some cases, filling a void that exists due to a lack of RCTs. To help facilitate these new initiatives, some researchers are taking advantage of the large, free database of deidentified data from over 40,000 patients who were cared for in the critical care units at Beth Israel Deaconess Medical Center (BIDMC) from 2001 to 2012 (Box 4.1). Called MIMIC-III Critical Care Database, it contains demographics, vital signs taken at the bedside, lab test findings, procedures, medications, and much more [25]. The database can provide source material for a variety of analytic studies. MIMIC-III and its earlier iterations have been used to fuel several published studies, as this truncated list demonstrates the following:

- A data-driven approach to optimized medication dosing: a focus on heparin (*Intensive Care Medicine*, 2015)
- Dynamic data during hypotensive episode improves mortality predictions among patients with sepsis and hypotension (*Critical Care Medicine*, 2013)
- Mortality prediction in intensive care units with the Super ICU Learner Algorithm: a population-based study. (*Lancet Respiratory Medicine*, 2015)
- A targeted real-time early warning score for septic shock. (*Science Translational Medicine*, 2015.)

Enthusiasm about the value of data analytics needs to be tempered with an understanding of its limitations, however. Analyzing data from EHRs, social media, mobile apps, remote sensors, medical claims, and other "real-world" evidence usually generates correlations between variables, not cause and effect relationships. It may reveal a strong temporal relationship between a diet rich in fruits and vegetables and a reduced incidence of cardiovascular disease (CVD), but it does not firmly establish the theory that such diets will prevent or effectively treat CVD. In this scenario, however, there is no harm in prescribing the dietary regimen to patients at risk of CVD since there are no adverse effects, and since there are several other lines of evidence suggesting a cause and effect relationship.

On the other hand, an analysis of thousands of patient records that finds a temporal relationship between statins, for example, and a reduced incidence

Box 4.1 Data Analytics at Beth Israel Deaconess Medical Center [27]

Since the 1970s, BIDMC has been at the forefront of medical informatics and it now deeply involved in conducting the type of data analytics that informs patient care. The overarching focus of BIDMC's Big Data initiatives is straightforward: Health-care data are useless unless they are used in the service of patients; and for that to happen, data must be transformed into information, information into knowledge, and knowledge into wisdom. This transformation occurs through CDSS, predictive modeling, a clinical query tool, and data mining of unstructured clinical content.

One of the primary goals of CDSS at BIDMC is to apply filters to the pages and pages of clinical and administrative data on each patient so that clinicians are not overwhelmed with more data than they have time to consume. CDSS offer up succinct suggestions to turn the data into actionable wisdom. For example, if a clinician orders a chest radiograph, a query is sent to the CDSS, which then compares the order to the best practice from the American Academy of Radiology and the medical literature. This expert advice is combined with data from the EHR about the patient's medication, lab results, and other relevant information to provide the physician with the optimal radiograph recommendation.

While this type of clinical support is invaluable in individualizing patient care, in today's health-care environment, with its emphasis on cost containment, health-care systems have to do more, becoming accountable for the health and wellness of large patient populations and monitoring the status of individual patients over long periods of time. Like several other large health systems, BIDMC is now using predictive modeling to help accomplish that goal, analyzing at-risk patients and looking for disease patterns. The intelligence derived from such predictive analytics is used to help home health care nurses catch emerging problems before they balloon out of control and require ED visits and expensive inpatient stays.

BIDMC, one of the hospitals affiliated with Harvard Medical School, is also deeply involved in clinical research. With that in mind, the hospital needs to mine 7 petabytes of data in help design research studies, including comparative effectiveness research. The medical center created Clinical Query to facilitate these efforts (see Fig. 4.2).

CQ2, the second iteration of this tool, lets clinicians and investigators search through clinical data that have been stored in BIDMCs database, which consists of millions of EHR records for patients treated at affiliated hospitals and medical practices. These records provide details on a wide variety of diagnoses, treatment plans, problem lists, allergies, and so on. Using CQ2, it is possible to gauge the merits of launching a clinical trial. For example, if there is a plausible

Continued

Box 4.1 Data Analytics at Beth Israel Deaconess Medical Center [27]—cont'd

Clinical Query 2 Web Interface

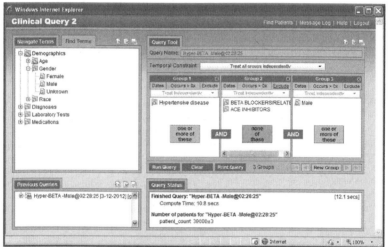

Figure 4.2 Clinical Query lets clinicians and researchers search through clinical data stored in BIDMC's database, which consists of millions of EHR records for patients treated at affiliated hospitals and medical practices. *BIDMC*, Beth Israel Deaconess Medical Center; *EHR*, electronic health record.

mechanism of action to suggest that high dose nonsteroidal antiinflammatory drugs contribute to liver cancer, it is possible to query the database to obtain the records of all the patients that were prescribed the drugs and who subsequently developed liver cancer. The records, which are deidentified to protect patients' privacy, can then be used as a jumping-off point to design a new clinical trial or retrospective data analysis.

Although CQ2 is an invaluable tool for clinicians and researchers, it has its limitations, one of which is its inability to extract anything useful from all the unstructured data located in EHRs, including discharge summaries, operative notes, and the like. For that, BIDMC turns to third party software that can perform NLP. NLP can convert long-winded narratives into short, structured data points that include computer-friendly SNOMED-CT or RxNorm terms. The information, knowledge, and wisdom generated with the help of NLP are used by BIDMC's case management specialists to improve patient care.

of colorectal cancer does not necessarily warrant prescribing these drugs to patients with the cancer, especially in light of their side effects, and the fact that there are drugs that have already been proven effective against the malignancy.

Whether or not a data analysis that detects a correlation is actionable or not, then, will depend on the evidential context. An analysis that detects a treatment option when an RCT-confirmed protocol does not exist should be given more weight than an analysis in which clinical trials have established a treatment protocol. And an analysis that individualizes the results of a RCT, as was the case with the Sussman study, should be given more weight that the RCT alone.

Finally efforts are being made to incorporate real-world data into clinical research. Rachel Sherman, MD, MPH and her colleagues describe these initiatives [26] as follows:

The technological and methodologic challenges presented by these new data sources are the focus of active efforts by researchers. For example, multiple stakeholders, including the Food and Drug Administration (FDA), are working on ways to harmonize data collected from EHRs, claims data, and registries to create a unified system for monitoring the safety and effectiveness of medical devices. Others, such as the National Institutes of Health (NIH) Collaboratory (an NIH Common Fund initiative devoted to building infrastructure, operational knowledge, and capacity for pragmatic research in the context of health care systems), are developing and implementing methods for incorporating data from EHRs and other sources into research. Such efforts include the development of large-scale distributed research networks and "computable phenotypes" (i.e., conditions or patient characteristics that can be derived from EHRs and claims data without requiring external review or interpretation) that allow researchers to identify cohorts of interest across multiple data sources.

REFERENCES

[1] Stolley PD, Lasky T. Investigating disease patterns: the science of epidemiology. New York: W.H.H. Freemanm; 1998. p. 36.

[2] Hurwitz A, Nugent A, Halper F, et al. Big data for dummies. Hoboken (NJ): John Wiley and Sons; 2013. p. 285.

[3] Obermeyer Z, Emanuel EJ. Predicting the future — big data, machine learning, and clinical medicine. N Engl J Med 2016;375:1216–9.

[4] Gulshan V, Peng L, Coram M, et al. Development and validation of a deep learning algorithm for detection of diabetic retinopathy in retinal fundus photographs. JAMA 2016;316(22):2402–10. Published online November 29, 2016.

[5] Graham DJ, Campen D, Hui R, et al. Risk of acute myocardial infarction and sudden cardiac death in patients treated with cyclo-oxygenase 2 selective and non-selective non-steroidal anti-inflammatory drugs: nested case-control study. Lancet 2005;365:475–581.

[6] Government Accountability Project. Dr. David Graham's full story. 2017. https://www.whistleblower.org/node/743.

[7] Sussman JB, Kent DM, Nelson JP, et al. Improving diabetes prevention with benefit based tailored treatment: risk based reanalysis of Diabetes Prevention Program. BMJ 2015;350:h454. http://www.bmj.com/content/350/bmj.h454.

[8] Kent DM, Hayward RA. Limitations of applying summary results of clinical trials to individual patients. JAMA 2007;298:1209–12.

[9] Ioannidis JP, Lau J. Heterogeneity of the baseline risk within patient populations of clinical trials: a proposed evaluation algorithm. Am J Epidemiol 1998;148(11):1117–26.

[10] Rothwell PM, Warlow CP. Prediction of benefit from carotid endarterectomy in individual patients: a risk-modelling study. European Carotid Surgery Trialists' Collaborative Group. Lancet 1999;353(9170):2105–10.

[11] Morrow DA, Antman EM, Snapinn SM, et al. An integrated clinical approach to predicting the benefit of tirofiban in non-ST-elevation acute coronary syndromes: application of the TIMI risk score for UA/NSTEMI in PRISM-PLUS. Eur Heart J 2002;23(3):223–9.

[12] Thune JJ, Hoefsten DE, Lindholm MG, et al. Simple risk stratification at admission to identify patients with reduced mortality from primary angioplasty. Circulation 2005;112(13):2017–21.

[13] Escobar GJ, Puopolo KM, Wi S, et al. Stratification of risk of early-onset sepsis in newborns ≥34 weeks' gestation. Pediatrics 2014;133:30–6.

[14] Centers for Medicare and Medicare Services. Readmissions reduction program (HRRP). April 18, 2016. https://www.cms.gov/Medicare/Medicare-Fee-for-Service-Payment/AcuteInpatientPPS/Readmissions-Reduction-Program.html.

[15] Amarasingham R, Patel PC, Toto K, et al. Allocating scarce resources in real-time to reduce heart failure readmissions: a prospective, controlled study. BMI Qual Saf 2013;22:998–1005.

[16] Freiman JE, Chalmers TC, Smith H, et al. The importance of beta, the type II error and sample size in the design and interpretation of the randomized control trial. N Engl J Med 1978;299:690–4.

[17] Moher D, Dulberg CS, Wells GA. Statistical power, sample size, and their reporting in randomized controlled trials. JAMA 1994;272:122–4.

[18] Tu W, Decker BS, He Z, et al. Triamterene enhances the blood pressure lowering effect of hydrochlorothiazide in patients with hypertension. J Gen Intern Med 2016;31(1):30–6.

[19] Simonsen L, Gog JR, Olson D, et al. Infectious disease surveillance in the big data era: towards faster and locally relevant systems. J Infect Dis 2016;214(Suppl. 4):S380–5.

[20] Viboud C, Charu V, Olson D, et al. Demonstrating the use of high-volume electronic medical claims data to monitor local and regional influenza activity in the US. PLoS One July 29, 2014;9(7):e102429. http://journals.plos.org/plosone/article?id=10.1371/journal.pone.0102429.

[21] Centers for Disease Control and Prevention. Flu activity forecasting website launched. January 19, 2016. http://www.cdc.gov/flu/news/flu-forecast-website-launched.htm.

[22] Parikh RB, Kakad M, Bates DW. Integrating predictive analytics into high-value care the dawn of precision delivery. JAMA 2016;315:651–2.

[23] Fihn SD, Francis J, Clancy S, et al. Insights from advanced analytics at the veterans health administration. Health Aff 2014;33:1203–11.

[24] Bombardier C, Laine L, Reicin A, et al. Comparison of upper gastrointestinal toxicity of rofecoxib and naproxen in patients with rheumatoid arthritis. N Eng J Med 2000;343:1520–8.

[25] MIMIC-III critical care database. https://mimic.physionet.org/about/mimic/.

[26] Sherman RE, Anderson SA, Dal Pan GJ, et al. Real-world evidence — what is it and what can it tell us? N Engl J Med 2016;375:2293–7.

[27] Halamka J. Early experiences with Big data at an academic medical center. Health Aff 2014;33(7):1132–8.

CHAPTER FIVE

How Mobile Technology and EHRs Can Personalize Healthcare

Although the theme of the book is realizing the promise of precision/personalized medicine, that theme does not imply that clinicians are not currently providing individualized care. Every time a physician substitutes one antibiotic for another that is causing adverse effects, they are practicing personalized medicine. Whenever a clinician chooses one insulin formula over another to fit an individual patient's lifestyle, they are taking a precise approach to patient care. The list of such decisions made to tailor medical care to each patient could easily fill this entire chapter. But with the introduction of new digital tools not traditionally in use, patient care can become even *more* personalized.

Mobile technology has the potential to personalize patient care by providing phone apps that give clinicians more detailed physiological parameters to evaluate, by linking patients and their practitioners via telemedicine services and text messaging, and by means of remote sensors that track metrics that would normally be missed. Similarly, electronic health records can record detailed genomic and environmental data that can contribute to each patient's condition. But like most other tools, these have their strengths and weaknesses, which we will explore below.

ACCESSING PATIENTS' "OTHER" DATA

"So much of what determines a person's health and well-being is independent of medical care." That observation, voiced by Rishi Sikka, a VP with Advocate Health Care in Illinois, is one of the reasons why mobile technology offers so many possible ways to personalize care [1]. It also implies that collecting blood pressure readings, serum cholesterol levels, and numerous other types of medical data during an office visit provides clinicians with a very limited view of what is really going on in a patient's body. And since the average patient spends perhaps 30 min in a doctor's office in a month and the remaining 700+ hours elsewhere, having insights into their

Realizing the Promise of Precision Medicine
ISBN 978-0-12-811635-7
http://dx.doi.org/10.1016/B978-0-12-811635-7.00005-1

behavior and health status during those hours can have a significant impact on treatment choices.

Numerous mobile apps have been developed to collect physiological and psychosocial metrics that can help practitioners improve disease management and boost clinical outcomes. But this field has become so crowded that it is difficult for clinicians and patients to determine which digital tools have real merit and which are mostly "smoke and mirrors." Few mobile apps have undergone the type of rigorous study that would make most clinicians comfortable using them, but there are many exceptions.

Kevin Cook, MD, from the Division of Allergy, Asthma, and Immunology at Scripps Clinic in San Diego, and his colleagues designed a smartphone app to individualize care for patients with asthma, using the guidelines from the National Asthma Education and Prevention Program to help them create the digital tool. The Scripps Clinic app, which used the URXmobile System platform, lets patients receive alerts and allowed them to interact with the program by submitting information inquiries and by entering relevant data. The interaction generated individualized coaching that helped patients self-manage their asthma.

Sixty patients with poorly controlled asthma were recruited for this proof of concept study and filled out four Asthma Control Test (ACT) surveys to help them and their clinicians evaluate shortness of breath, nocturnal symptoms, use of rescue inhalers, overall feelings of disease control, and the impact that the disease had on their ability to carry out daily activities over a period [2]. Cook and his associates found that patients' scores on the ACT surveys increased from 16.6, which represented inadequate to poor asthma control to 20.5, which indicated a controlled state of affairs.

One of the potential limitations of developing a more personalized approach to patient care is that it may require a much greater time commitment from clinicians. It is important to note that the Scripps Clinic asthma app may actually save clinicians time. Because the app continuously collected patient data, including current medication, peak flow values, and their usage of educational materials, it was able to dynamically analyze each patient's situation, detect worrisome trends, and then create individualized interventions and proactive alerts based on the National Asthma Education and Prevention Program treatment guidelines. Patients who scored a low rating on the ACTs received an alert to encourage them to make changes. Those who continued to exhibit poor control were given the clinic phone number and their physician's email address to follow up.

Although Cook et al.'s results were positive, a recent Cochrane review of several studies that evaluated home telemonitoring of patients with asthma between clinic visits was less enthusiastic [3]. The analysis looked at 18 studies that included over 2200 participants with mild to moderate persistent asthma; the studies followed patients for up to 12 months. It concluded that there was no clear evidence that telemonitoring, which included feedback from clinicians, increased or decreased the likelihood of exacerbations requiring oral steroids or a hospital stay. The report did, however, suggest a possible improvement in patients' quality of life and lung function, when compared to patients who did not have telemonitoring.

THE ROLE OF MOBILE TECHNOLOGY IN DIABETES CONTROL

Of all the disorders that may respond to the individualized care afforded by mobile apps, diabetes mellitus likely tops the list. One of the reasons the disease is such a good fit for mobile technology is its very nature of demanding a great deal of self-management, in contrast to acute conditions such as appendicitis or abdominal hernia, for which the clinician plays the dominant role. (Patients who attempt self-appendectomy usually do not fare very well.)

There is ample evidence to show that diabetes self-management combined with diabetes education is a cost-effective way to lower hemoglobin A1c (HbA1c) readings, the best metric we have to monitor a patient's glucose control [4]. What remains unresolved is how much of an impact mobile apps and other computer-assisted protocols have on improving diabetes self-management and clinical outcomes.

One review of 16 randomized controlled trials that recruited over 3500 patients with Type 2 diabetes found that computer-assisted self-management levels lowered HbA1c by 0.2%. The same analysis found that smartphone-based programs in particular lowered HbA1c levels by 0.5% [5]. To put those findings into context, consider an analysis of over 4500 patients from 23 hospital-based clinics, which found that "Any reduction in HbA1c is likely to reduce the risk of complications, with the lowest risk being in those with HbA1c values in the normal range (<6.0%)." More specifically, a 15% drop in HbA1c was associated with a 37% decrease in the risk of microvascular complications and a 21% drop in diabetes-related death [6]. A decrease from about 9.5% to 9%, the equivalent of a 0.5% drop, was associated with a decrease in the adjusted incidence of microvascular complications from ~100 per 1000 person years to ~75 per 1000 years [6,7].

Some studies, however, have suggested that having Type 2 patients check their own blood glucose with a home monitor is not especially cost-effective, especially if they are not on insulin therapy [8,9]. On the other hand, researchers who asked Type 1 diabetic patients to attach a device (iBGStar) that can send blood glucose readings to an iPhone reported better glycemic control according to Shah and associates [10].

Similarly, positive data have been published to demonstrate that a system like the BlueStar mobile app, which provides personalized behavioral intervention to Type 2 diabetics, can improve the interaction between patient and provider and reduce HbA1c levels over 1 year. The coaching app reduced these levels by 1.9% in the group in the maximum treatment arm, compared to only 0.7% in the group receiving usual care [11]. Maximal treatment included the use of the mobile app, web-based self-management, and provider assistance. Unlike many mobile diabetes apps, BlueStar has received Food and Drug Administration (FDA) clearance as a medical device and must be prescribed by a health professional. The reports generated by the software are sophisticated enough to serve as clinical decision support (CDS) to tailor each patient's treatment plan.

In a separate evaluation, Sarah Wild, with the Usher Institute of Population Health Sciences and Informatics, University of Edinburgh, and colleagues recently tested the value of telemonitoring in patients with poorly controlled Type 2 diabetes [12]. Patients were asked to measure their blood glucose levels at least twice weekly, including a fasting and one nonfasting reading. They were instructed to transmit those readings via Bluetooth technology through a supplied modem to a remote server that was monitored by nurses. Over a 9-month intervention period, the nurses checked the data weekly to make changes in patients' treatment regimen when warranted, based on national guidelines. When HbA1c data were tabulated for 146 patients on the experimental program and 139 controls, the researchers discovered that the readings declined by a clinically significant 0.51% in the telemonitored patients (63 vs. 67.8 mmol/mol, or 8.9% vs. 9.4%).

It is important to note, however, that the success of this particular program did require clinician support, which would not be feasible if a medical practice did not have the resources to employ qualified nurses to redirect treatment choices. But it is worth mentioning that the 95% confidence interval for the percentage change in HbA1c was 0.22%–0.81%, which implies that some patients saw much better results than others. The mixed results from all aforementioned studies also suggest that clinicians need to choose a mobile solution that fits each individual's needs and preferences.

One mobile solution worth considering for patients on insulin therapy is continuous glucose monitoring (CGM). At least three companies—Dexcom, Medtronic, and Abbott—make systems that include a beneath-the-skin sensor, transmitter, and wireless monitor that continuously records blood glucose levels and generates detailed trending data (see Fig. 5.1).

In 2015, the FDA approved the first mobile medical app platform for CGM, created by Dexcom Inc. The Dexcom Share Direct lets patients share real-time data through an Apple device like an iPhone. The system is especially helpful for parents who want to remotely monitor their child's glucose levels. According to FDA, "The Dexcom Share system displays data from the G4 Platinum CGM System using two apps: one installed on the patient's mobile device and one installed on the mobile device of another person. Using Dexcom Share's mobile medical app, the user can designate people ('followers') with whom to share their CGM data. The app receives real-time CGM data directly from the G4 Platinum System CGM receiver and transmits it to a Web-based storage location. The app of the 'follower' can then download the CGM data and display it in real-time." [13]

Systems like this can help personalize diabetes care in two respects. The fact that they provide clinicians, families, and patients more detailed information on blood glucose readings over time allows them to fine-tune insulin therapy, food intake, and activity levels. But the settings on the devices themselves can also be personalized by dialing in specific alarm thresholds for hypo- and hyperglycemia.

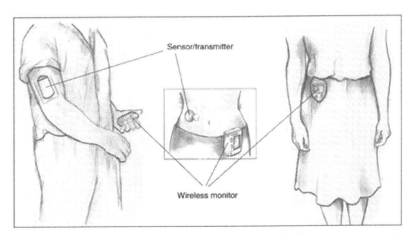

Figure 5.1 Continuous glucose monitoring systems provide glucose measurements as often as once per minute. The measurements are transmitted to a wireless monitor. *(NIH National Institute of Diabetes and Digestive and Kidney Diseases.)*

The American Association of Clinical Endocrinologists and the American College of Endocrinology recommend CGM for all adult patients with Type 1 diabetes and in patients with Type 2 diabetes who are required to take several insulin injections daily, and those on basal insulin or sulfonyl-ureas (a type of oral medication that lowers blood glucose) who are likely to be unaware of impending hypo- or hyperglycemia. Similarly the American Diabetes Association recommends Type 1 patients who experience hypoglycemia unawareness or experience frequent episodes of hypoglycemia also use CGM [14].

Lastly, a recent randomized controlled study conducted by the well-respected Joslin Diabetes Center in Boston is worth consideration [15]. As most clinicians who care for Type 2 diabetic patients know, one of the most difficult periods for patients is when they are required to add insulin to a treatment regimen that includes an oral agent like metformin. Titrating the dose is challenging for patients and physicians alike, and incorrect insulin use remains one of the main reasons patients end up in the emergency room. The experts at Joslin have devised a creative way to utilize mobile technology to assist with this transition, easing the burden on clinicians and empowering individual patients to take on more of the responsibility for adjusting their basal insulin dose.

William Hsu, MD, and his colleagues designed the system using a software platform from Massachusetts Institute of Technology's Media Lab called CollaboRhythm. The cloud-based diabetes management system makes use of an individualized treatment plan agreed on between clinician and patient, a plan that patients access on their tablet. A wireless glucose meter is patched into the system to transmit blood glucose readings.

Their program is unique in that it incorporates situated learning theory to help patients learn to be relatively self-sufficient in managing their condition. The clinician explains their decision-making process for choosing an insulin dose, and then encourages the patient to develop the same skill by active participation rather than through lectures or handouts, gradually developing the expertise needed to adjust their dose. The study, which took place over an average of 12 weeks and included 40 patients, was able to lower HbA1c levels by 3.1%, compared to 2% in the control group. Equally important was the fact that practitioners were able to spend less time with the experimental group, about 66 versus 82 min/patient. Although their approach is probably too complex to incorporate into a primary care practice, it is certainly worth considering in a busy specialty practice. It may ease the time commitments for clinicians while giving committed patients a sense of self-empowerment.

THE ROLE OF MOBILE TECHNOLOGY IN OTHER DISEASES

There is evidence to suggest that text messaging, health apps, and other digital tools can personalize care for patients with coronary heart disease and other chronic disorders, as well as improve clinical outcomes. For example, the Tobacco, Exercise, and Diet Messages or TEXT ME trial, a large single-blind randomized study, found that texting semipersonalized health messages to patients with preexisting heart disease reduced low density lipoprotein (LDL) cholesterol levels by 5 mg/dL and systolic blood pressure by 7.6 mmHg after 6 months. It also increased physical activity by 345 MET (metabolic equivalents) minutes/week, and significantly reduced smoking ($P < .001$) [16]. And although the researchers who conducted this trial refer to their results as a "modest improvement in LDL cholesterol," previous studies have demonstrated that every 1% drop in LDL level results in a 1% reduction in coronary heart disease deaths and nonfatal myocardial infarction [17].

Unlike the program spearheaded by Sarah Wild and her colleagues, this text messaging experiment involved less provider time and energy. The automated electronic delivery system sent out four messages a week for 24 weeks and used an algorithm that was able to personally address some messages and personalize the content of the message based on each patient's baseline characteristics. Some messages, for instance, focused on smoking if the subject used tobacco. One of the limitations of the studies, however, was its short-term nature. It was not possible to determine if the effects the text messages had on behavior and the clinical benefits would last over time. This is an important issue because mobile health apps seem to engage many patients because they are a relatively new way to interact with their healthcare provider. When the novelty wears off, patients may lose interest and see diminishing returns.

The American Heart Association (AHA) has done an exhaustive review of the research on the role of mobile technology, concentrating on the prevention of cardiovascular disease. AHA focused on how mobile technology might influence the risk factors most likely to contribute to heart disease, including obesity, lack of physical exercise, hypertension, and elevated cholesterol levels [18]. The results were mixed, which suggests that these tools work in some populations and not others and different types of mobile interventions produce different results.

Five of eight randomized controls trials conducted in the United States found significantly more weight loss occurring in patients using mobile technology than in controls. Many of the positive studies used text messaging or text messaging plus Facebook to encourage weight loss.

For clinicians and technologists searching for a common demonstrator for success among mobile weight loss tools, an analysis by Khaylis and associates is useful. They identified five features that have been linked to effective technology-based weight loss programs [19]:

- A structured approach
- Self-monitoring
- Feedback and communication
- Social support
- The ability to individualize the intervention

Elsewhere in its review, AHA concludes that:

Mobile interventions can produce weight loss in motivated populations, albeit at a lower magnitude relative to traditional treatment approaches. The characteristics of successful mobile interventions are quite comparable to those of their offline counterparts: The largest weight losses are produced by comprehensive, multicomponent interventions that are personally tailored, promote regular self-monitoring, and involve a qualified interventionist. The accumulated evidence, although limited, supports intervention delivery through a range of technology channels (including the Web, SMS, e-mail, telephone, and IVR), with limited variability in the magnitude of weight loss outcomes.

But on a cautionary note, it also states that although the evidence strongly supports the value of texting when supported by phone calls, websites, and social media, "there is no evidence to suggest that SMSs as a stand-alone intervention are effective."

Finally, as providers consider using mobile technology, it is important to remember that weight loss programs do not have to result in a huge weight reduction to have benefits. AHA points out that a sustained weight loss of only 3%–5% can have an impact, significantly reducing the risk of cardiovascular disease. In practical terms, that means a 250 pound person only needs to lose between 7.5 and 12.5 pounds to see results.

Similarly when evaluating mobile-enhanced programs that encourage exercise, there is no need to reach for the moon. The Centers for Disease Control and Prevention recommends at least 30 min of moderate intensity physical activity on most days of the week. And AHA reminds readers that sustained exercise reduces the risk of Type 2 diabetes, stroke, osteoporosis, and depression. It also reduces the risk of cardiovascular disease by lowering blood pressure. AHA reported the results of 14 RCTs it considered of high quality, which used texting on a mobile phone, email, a pedometer, and the Internet. Nine of the 14 reported the programs were effective in increasing physical activity.

The use of mHealth interventions to help patients gain control over their hypertension also may have merit, but the data are less robust. There is ample evidence to demonstrate that self-measured blood pressuring monitoring (SMBP) is valuable in managing hypertension, which is why the Seventh Report of the Joint National Committee on Prevention, Detection, Evaluation, and Treatment of High Blood Pressure recommended its use. But while SMBP monitoring, strictly speaking, is considered a mobile technology, it's not exactly the cutting edge. To date, however, research on more sophisticated digital tools has suffered from one major weakness. Most high-quality studies have only lasted 6 months or less. Since hypertension is a chronic disease requiring long-term solutions, we need to establish the effect of mobile technology over many years.

Despite this shortcoming, it is worth mentioning that five high-quality studies using SMBP plus support found a reduction in systolic blood pressure of 2.1 and 8.3 mmHg. (Only one of the studies included mHealth support.) To put these numbers into perspective: a 1 mm Hg drop in systolic blood pressure has been associated with about 20 fewer cases of heart failure (HF) per 100,000 person-years in African Americans and about 13 fewer HF deaths in Whites [20].

There have also been some attempts to enlist the help of mobile technology to manage dyslipidemia, with promising but tentative results. There are home lipid testing kits for use with a smartphone for instance, but the 2015 AHA guidelines concluded that "the amount of evidence-based literature in this area remains surprisingly low."

In addition to mobile technology designed to address weight gain, hypertension, and other individual risk factors for heart disease, there are also applications that measure one's overall risk. The Marshfield Clinic's HeartHealth Mobile app is a well-respected tool in this category. It is available for Apple devices and as a web version also exists. Users can insert statistics on their height, weight, cholesterol level, smoking status, the presence of diabetes, HbA1c level, and blood pressure and obtain a risk score. The app joins a growing number of digital tools that have been designed with gamification principles in mind with the hope that it will keep users entertained while it improves their health.

One area of concern for persons at risk of heart disease that was not addressed by the AHA mobile technology review in any depth is emotional health. The evidence linking psychosocial stress to heart disease is undeniable. And new data from the long-running INTERHEART study only support the relationship. Analysis of over 12,000 cases of acute myocardial infarction (AMI), conducted in 52 countries, has found that the odds of having an AMI are three times greater among individuals who were angry and involved in heavy physical exertion within an hour of having the attack, when compared to AMI patients who did not share these risk factors [21]. The study paints a dramatic picture: Imagine a spouse arguing with his or her mate, boiling over with contempt and frustration, and then storming out of the house to relieve their pent-up emotions by vigorously shoveling snow or retreating to the gym to furiously run on a treadmill, only to fall victim to their own sympathetic nervous system and its effects on the myocardium. As Smyth and associates explain in their INTERHEART study report: "Physical exertion and emotions (including anger and emotional upset) are reported to cause sympathetic activation, catecholamine secretion, systemic vasoconstriction, and increase heart rate and blood pressure, thereby modifying myocardial oxygen demand, which may precipitate the rupture of an already vulnerable atherosclerotic plaque."

REMOTE PATIENT MONITORING

In light of such evidence, a precision medicine assessment would be incomplete without the use of a measuring stick to assess an individual's ability to manage stress and other psychosocial problems. The data from Smyth et al. further highlight the importance of such an assessment tool. While they found an odds ratio of 3.05, they also found that only 14.4% of the study population who were angry or emotionally upset experienced an AMI, which implies that nearly 85% did not. Identifying those who are most sensitive to extreme physical exertion and anger will require a sensitive screening system.

Brent Winslow from Design Interactive Inc, in conjunction with colleagues from the Philadelphia VA Medical Center and the University of Pennsylvania, has developed a wearable sensor that may have value in measuring a person's stress response and serve as a monitoring device. They used the Biopac MP-150 system to collect physiological data—including cardiovascular and electrodermal activity—both of which signal changes in a person's response to stress. To compare these readings against a standard parameter that indicates a stress reaction, the researchers also measured salivary cortisol levels.

The system was tested in a group of armed forces veterans who were undergoing cognitive behavioral therapy (CBT) for stress and anger management, and on healthy controls. The system was able to detect physiological stress in more than 90% of the controls, and veterans who used the system while undergoing therapy were significantly improved on measures of stress, anxiety, and anger, when compared to veterans who underwent CBT alone [22].

Wearable sensors are also finding a place in pediatric medicine. Children's of Alabama, located in Birmingham, is providing the parents of children with congenital heart disease with a remote monitoring kit from Vivify Health that includes a tablet and sensors to measure patients' weight, pulse, and oxygen saturation. The "Hearts at Home" program helps parents cope with the most vulnerable stages of postsurgery and recovery and gives clinicians immediate access to vital signs that may require prompt action [23].

Personalized patient care also means understanding which patients are responding to their medications and which ones are not. As we mentioned in a previous chapter, pharmacogenomics can play an important role in determining individual response to medication. But at a much more basic level, knowing which patients are actually *taking* their medication is just as important. There are numerous digital tools available that can help in this regard. They include smartphone apps such as MyMedSchedule, MyMeds, and RemindMe, which can remind patients to take their medication. Other tools include "smart pill bottles" that keep track of when patients open and close their pill vials. There is even a "smart necklace" called WearSens that contains a piezoelectric sensor that allows clinicians to monitor patients' medication adherence by detecting a person's neck movements when they swallow a pill [24].

Another area of patient care that has seen a great deal of interest in mobile technology is pressure ulcers. A systematic review of 36 high-quality research studies found a variety of sensor-based approaches that help clinicians monitor the risk of developing pressure sores in patients confined to bed. Most methods involved the use of sensors installed in mattresses but others made use of electronic sensors and tactile sensory coils. Sensors were used to measure temperature, pressure, the humidity of a patient's body, and blood flow. Thirteen of the articles cited in the review concluded that the technology was effective in controlling risk factors for pressure sores or were able to generate useful reports, including reports that informed clinicians of a patient's position [25]. None of the investigations were randomized or controlled, however.

With so many remote sensors available to measure a wide variety of physiological parameters and limited regulation on the industry generating these sensors, critics have rightly asked whether these products produce accurate readings. Robert Wang, MD, a cardiologist with the Cleveland Clinic, and his associates addressed this issue by testing the accuracy of heart rate monitoring available in wearables from Apple Watch, Fitbit Charge, Mio Alpha, and Basis Peak, comparing their readings to the heart rate recorded by a standard EKG machine as the gold standard. Volunteers were tested while at rest and while on a treadmill at 2, 3, 4, 5, and 6 mph. When they tested 50 healthy adults, they found the Apple Watch and Mio Fuse readings correlated closely to the EKG readings (concordance correlation coefficient .91 for each). The Fitbit and Basic Peak were less accurate (.84 and .83). While such variance is of little concern when the devices are used for recreational purposes, some clinicians are now using them to monitor cardiac patients in rehabilitation, situations in which accuracy is much more important. In the case of the Basis Peak device, the researchers found a median difference of -8.9 and -7.3 bpm at 2 and 3 mph [26]. And since this experiment included only healthy adults, it is entirely possible the results would have been worse in cardiac patients with damaged blood vessels.

Fitness apps that measure heart rate do not require FDA approval, and as such do not have to reach the high standard of the medical devices that have met FDA standard. Pulse oximeters, EKG monitors and several other remote sensing devices do have to meet this higher standard, which means clinicians can have more confidence in their reliability.

A pulse oximeter, for instance, must meet a series of criteria to gain FDA clearance, including accuracy testing, as explained in the Agency's guidelines [27]. Meeting this higher standard gives health insurers the confidence to recommend their use in home care, as illustrated in Aetna's stated policy [28]. It also offers a measure of assurance that using such devices can individualize patient care, at the same time increasing the likelihood that patients can be reimbursed for the cost of the equipment, under specified conditions. Home monitoring devices that have also met FDA criteria include the smartphone-based AliveCor EKG and the HeartCheck handheld EKG device. (For a summary of how the FDA determines which mobile technology requires its approval or clearance and which does not, see Box 5.1).

Box 5.1 When does a medical app require FDA clearance?

The FDA's position on most issues is rarely simple. Its guidelines on mobile health apps are no exception. The agency says it is taking a tailored, risk-based approach to the approval of mobile apps and that most health-related mobile apps will not come under scrutiny. It focuses primarily on those apps that fall under the category of medical devices, which in turn applies to two groups, those that are designed to act as accessories to regulated medical devices, and those that transform a mobile platform into a regulated medical device [38].

The FDA gives examples of the mobile apps that have required FDA clearance as medical devices. The list includes Accu-Chek's Connect Diabetes Management app, Withings blood pressure monitor, a vital signs patch system that detects arrhythmias, Pocketview ECG software, an ultrasound imaging system, and many more. A list of examples has been posted on the FDA website.

And then there are those apps that the Agency says are devices but in its eyes do not pose enough of a threat to public safety to warrant a closer look. In this third category are apps for which the FDA says it will "exercise enforcement discretion." In such cases, it will not expect manufacturers to submit premarket review applications or to register and list their apps with the FDA. Among this group are the following [39]:

- Mobile apps that help patients with diagnosed psychiatric conditions (e.g., posttraumatic stress disorder, depression, anxiety, obsessive compulsive disorder) maintain their behavioral coping skills by providing a "Skill of the Day" behavioral technique or audio messages that the user can access when experiencing increased anxiety;
- Mobile apps that provide periodic educational information, reminders, or motivational guidance to smokers trying to quit, patients recovering from addiction, or pregnant women;
- Mobile apps that use GPS location information to alert asthmatics of environmental conditions that may cause asthma symptoms or alert an addiction patient (substance abusers) when near a preidentified, high-risk location;
- Mobile apps that use video and video games to motivate patients to do their physical therapy exercises at home;
- Mobile apps that prompt a user to enter which herb and drug they would like to take concurrently and provide information about whether interactions have been seen in the literature and a summary of what type of interaction was reported;
- Mobile apps that help asthmatics track inhaler usage, asthma episodes experienced, location of user at the time of an attack, or environmental triggers of asthma attacks;
- Mobile apps that prompt the user to manually enter symptomatic, behavioral or environmental information, the specifics of which are predefined by a health-care provider, and store the information for later review;

Continued

Box 5.1 When does a medical app require FDA clearance?—cont'd

- Mobile apps that use patient characteristics such as age, sex, and behavioral risk factors to provide patient-specific screening, counseling, and preventive recommendations from well-known and established authorities;
- Mobile apps that use a checklist of common signs and symptoms to provide a list of possible medical conditions and advice on when to consult a health-care provider;
- Mobile apps that guide users through a questionnaire of signs and symptoms to provide a recommendation for the type of health-care facility most appropriate to their needs;
- Mobile apps that record the clinical conversation a clinician has with a patient and sends it (or a link) to the patient to access after the visit;
- Mobile apps that are intended to allow a user to initiate a prespecified nurse call or emergency call using broadband or cellular phone technology;
- Mobile apps that enable a patient or caregiver to create and send an alert or general emergency notification to first responders;
- Mobile apps that keep track of medications and provide user-configured reminders for improved medication adherence;
- Mobile apps that provide patients a portal into their own health information, such as access to information captured during a previous clinical visit or historical trending and comparison of vital signs (e.g., body temperature, heart rate, blood pressure, or respiratory rate);
- Mobile apps that aggregate and display trends in personal health incidents (e.g., hospitalization rates or alert notification rates);
- Mobile apps that allow a user to collect (electronically or manually entered) blood pressure data and share this data through email, track and trend it, or upload it to a personal or electronic health record;
- Mobile apps that provide oral health reminders or tracking tools for users with gum disease;
- Mobile apps that provide prediabetes patients with guidance or tools to help them develop better eating habits or increase physical activity;
- Mobile apps that display, at opportune times, images or other messages for a substance abuser who wants to stop addictive behavior;
- Mobile apps that are intended for individuals to log, record, track, evaluate, or make decisions or behavioral suggestions related to developing or maintaining general fitness, health, or wellness, such as those that:
  - Provide tools to promote or encourage healthy eating, exercise, weight loss, or other activities generally related to a healthy lifestyle or wellness;
  - Provide dietary logs, calorie counters, or make dietary suggestions;
  - Provide meal planners and recipes;

Box 5.1 When does a medical app require FDA clearance?—cont'd

- Track general daily activities or make exercise or posture suggestions;
- Track a normal baby's sleeping and feeding habits;
- Actively monitor and trend exercise activity;
- Help healthy people track the quantity or quality of their normal sleep patterns;
- Provide and track scores from mind-challenging games or generic "brain age" tests;
- Provide daily motivational tips (e.g., via text or other types of messaging) to reduce stress and promote a positive mental outlook;
- Use social gaming to encourage healthy lifestyle habits;
- Calculate calories burned in a workout.

- Mobile apps for providers that help track or manage patient immunizations by assessing the need for immunization, consent form, and immunization lot number;
- Mobile apps that provide drug–drug interactions and relevant safety information (side effects, drug interactions, active ingredient) as a report based on demographic data (age, gender), clinical information (current diagnosis), and current medications;
- Mobile apps that enable, during an encounter, health-care providers to access their patient's personal health record (health information) that is either hosted on a web-based or other platform;
- Mobile apps that are not intended for diagnostic image review (and include a persistent on-screen notice, such as "for informational purposes only and not intended for diagnostic use") are medical image communications device under 21 CFR 892.2020, product code LMB. Nondiagnostic uses could include: image display for multidisciplinary patient management meetings (e.g., rounds) or patient consultation. Medical image communications devices are Class I and do not require FDA 510(k) premarket notification. Possible product code LMD (21 CFR 892.2020);
- Mobile apps that allow a user to collect, log, track, and trend data such as blood glucose, blood pressure, heart rate, weight, or other data from a device to eventually share with a heath care provider, or upload it to an online (cloud) database, personal or electronic health record;
- Mobile apps that provide the surgeon with a list of recommended intraocular lens powers and recommended axis of implantation based on information inputted by the surgeon (e.g., anticipated surgically induced astigmatism, patient's axial length, and preoperative corneal astigmatism, etc.)

WHERE DO EHRs FIT IN?

In Chapter 2, we discussed the role of electronic health records in the federally sponsored Precision Medicine Initiative. We also mentioned the fact that most of the currently available EHR systems fall short of the requirements needed to be valuable in a precision medicine model of care because of their inability to fully manage genomic and psychosocial data. That is slowly changing.

In some situations, EHR systems have become more personalized by integrating them with new algorithms to help identify specific subgroups of at risk patients. For example, at Partners Healthcare in Boston, Shirley V. Wang and associates designed algorithms to help identify patients with atrial fibrillation who were at greater risk of stroke and major bleeding. The algorithms used EHR data, including problems lists, billing codes, laboratory data, prescription data, vital signs, and clinical notes. This EHR-enhanced approach to personalized care was compared to a gold standard, namely manual chart reviews performed by physicians. The algorithms successfully detected atrial fibrillation patients at risk for 11 conditions, with a median positive predictive value greater than 0.9 [29].

Being able to identify high-risk patients with atrial fibrillation has important treatment implications. Oral anticoagulant therapy has been shown to reduce the risk of ischemic stroke by 68% annually among patients with nonvalvular atrial fibrillation. But these drugs also increase the risk of bleeding complications. Thus the ability to identify patients whose risk of stroke is high enough to warrant drug therapy can save lives. Unfortunately, while Wang and her colleagues were able to demonstrate that EHR data can help individualize care for atrial fibrillation patients, it is unlikely that data would be as useful without their specially designed algorithm. This point is summed up in their report: "One of the problems with using EHRs to improve care quality has been that the variables needed to measure and improve care quality may not yet be represented in EHRs in ways that make these data easy to leverage for these purposes." In other words, much of the information needed to justify prescribing anticoagulants already resides in EHRs, but it is so scattered or obscure that busy clinicians may not connect the dots.

Partners Healthcare has also been pioneering the integration of genomic data into patients' health records, a critically important function in the emerging world of precision medicine [30]. Partners uses the Epic EHR system and has implemented a common record system called eCare based on Epic. It allows patients and physicians at any of the 12 Partners Hospitals and all its medical practices to share data regardless of their location within the system. Equally important, Partners has devised a way to deliver genomic data to clinicians through an EHR plug-in called GeneInsight, thereby linking molecular laboratories to the records system. GeneInsight was recently sold to Sunquest Information Systems, which allows academic medical centers through the United States to use the service.

To fully appreciate the role of EHRs in realizing the promise of precision medicine, it helps to envision precision as a large ecosystem. The analogy is worth exploring. In nature, ecosystems consist of fauna and flora, climatic characteristics, soil conditions, geologic features, and a host of other interacting influences. Similarly, the precision medicine ecosystem is made of many interacting components, including patients, clinicians, researchers, laboratory services, CDS software, genomic databases, smartphones, servers, claims data, mobile apps, biobanks to store clinical specimens, and EHRs. EHRs need to serve as gateways to this ecosystem. And for the EHR to become an effective conduit, it needs a way to organize these diverse sources in a way that lets clinicians and patients make more effective diagnostic and treatment decisions.

The current population-based medical ecosystem makes use of many of the same components but remains dysfunctional in many respects. Medications are being prescribed for patients whose genetic makeup contraindicates their use; laboratory tests performed by one facility are often unavailable to clinicians in another location to facilitate an accurate diagnosis. And insights that can be gleaned from remote sensors are often never factored into the equation. In some respects, it is like our current natural ecosystem, where climate changes disrupt ocean levels, species are allowed to become extinct, and other species are artificially introduced into a community, all of which disrupts the balance of nature.

The emerging precision medicine ecosystem hopes to optimize patient care by improving on healthcare's infrastructure and by facilitating the accurate interpretation of data. As Samuel J. Aronson and Heidi L. Rehm [31] from Partners Healthcare Personalized Medicine explain the issue:

This ecosystem is beginning to link clinicians, laboratories, research enterprises and clinical-information-system developers together in new ways. There is increasing hope that these efforts will create the foundation of a continuously learning healthcare system that is capable of fundamentally accelerating the advance of precision-medicine techniques. Interpretation is key to the precision-medicine ecosystem. It occurs at several levels. Individual variants can be interpreted in relation to specific indications. Sets of variants can be assessed in relation to their collective impact on patients. Genetic and clinical data can be combined to determine the best course of action for a patient. The quality of these interpretations is highly dependent on the data on which they are based. For this reason, research and clinical databases provide the foundation for precision medicine. Continuous learning in health care is in many ways driven by improvements to the content and structure of these resources.

And while the quality of these interpretations is dependent on the quality of the data being generated, it is also dependent on an electronic health record system that can effectively deliver that data to clinicians at the point of care. It is also dependent on a CDS tool that can alert clinicians to well-known drug interaction or allergies, and one that can detect *patterns* in the massive amounts of data that are becoming available.

CLINICAL DECISION SUPPORT SYSTEMS

Throughout this book, we have used the terms precise medicine and personalized medicine interchangeably because they do in fact have very similar meanings, both implying a type of patient care that takes into account an individual's unique characteristics, including genetic makeup, psychosocial influences, exposure to environmental toxins, and so on. But there is another connotation to the word precision. Precision also implies exactness or accuracy, and it is this accuracy that CDS systems seek to achieve.

There is little doubt that clinicians need better CDS. A recent report from the Institute of Medicine (IoM), entitled *Improving Diagnosis in Health Care*, concluded that at least 5% of US adults experience a diagnostic error annually in an outpatient setting. Postmortem examination has also revealed diagnostic errors in about one in 10 patients, according to IoM, and diagnostic mistakes cause as many of 17% of hospital adverse events. The report concludes that "Diagnostic errors are the leading type of paid medical malpractice claims, are almost twice as likely to have resulted in the patient's death compared to other claims, and represent the highest proportion of total payments." [32] (Fig. 5.2).

Diagnostic mistakes occur for a variety of reasons, including the short duration of the typical office visit, the rapid pace at which clinicians are forced to perform, the complex differential diagnostic reasoning process required to detect uncommon diseases, and the massive amounts of data in the scientific literature that one must master to detect these less common disorders. (The latter problem is partially solved by CDS programs such as Watson Health, discussed later.)

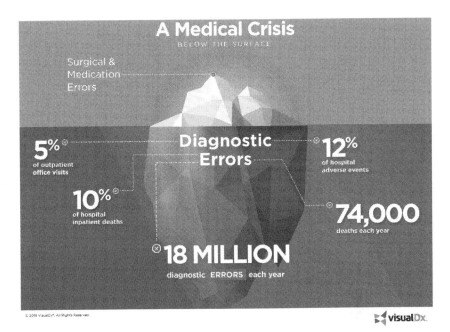

Figure 5.2 Medical diagnoses remain a major concern for hospitals and clinicians, a problem that clinical decision support systems are designed to address. *(Reprinted with permission from VisualDx.)*

Another potential obstacle to solving the misdiagnosis dilemma is the normal functioning of the human brain. Daniel Kahneman, a psychologist who won the Nobel Prize in Economics in 2002, postulates that the brain uses two systems during the reasoning process. System 1 is fast and initiative, relying on pattern recognition and memory, while System 2 is slower and more deliberate. The former is typically used by clinicians when routine decisions and familiar disorders present themselves. System 2 should take precedence when unexpected, challenging cases surface. But the shift from System 1 to System 2 analysis takes much more mental effort and time, which is why clinicians do not always make the necessary shift. A well-designed clinical decision support system (CDSS) can facilitate the switch from System 1 to System 2.

CDS software also has an important role in precision medicine because physicians are prone to several cognitive errors during the diagnostic process, including availability bias and attribution errors, to name a few. An analysis by Mark Graber and associates in *JAMA Internal Medicine*, for example, suggests that approximately three out of four diagnostic errors that occur in internal medicine are the result of cognitive errors. They concluded that: "Premature closure, i.e., the failure to continue considering reasonable alternatives after an initial diagnosis was reached, was the single most common cause. Other common causes included faulty context generation, misjudging the salience of findings, faulty perception, and errors arising from the use of heuristics." [33]

CDS systems come in a variety of configurations, some primitive and some quite sophisticated, some for specialized audiences and some for general practitioners. Ideally these decision-making tools should help practitioners circumnavigate some of the aforementioned cognitive errors and allow them to switch back and forth between System 1 and System 2 thinking. US Department of Health and Human Services has been supporting the use of CDS systems for several years. Its guide to integrating these tools into EHRs provides a blueprint for developers and end users to help judge their effectiveness. Based on the work initially done by Healthcare Information and Management Systems Society, it has categorized the components of existing CDS tools into six categories [34]:

1. Documentation forms and templates
2. Relevant data presentation
3. Order and prescription facilitators
4. Protocol pathway support
5. Reference information and guidance
6. Alerts and reminders

Some of these categories work well for clinicians in fast thinking mode, whereas others are useful when they must slow down. For fast thinking situations, chart note templates often work well. A simple condition such as uncomplicated upper respiratory infections, for example, does not require complex diagnostic reasoning or extensive narrative notes.

Feature 2, presenting relevant data in a way that clinicians can easily use, is more challenging, and when designed properly can facilitate slow and fast thinking situations. When CDS programs have the ability to present this data visually in charts, dashboards, and flow sheets, it can make diagnostic and treatment decisions more precise. These visualization tools often make clinical problems and troubling patterns easy to detect. Fig. 5.3 shows a simple example of how data-generated graphs in a CDS tool can make patient management more effective. By charting a patient's weight from 2001 to 2013, one can easily spot clinically relevant trends. In this case, the chart quickly tells a practitioner that from December 2002 to February 2009, the patient has gone from about 183 pounds to about 217 pounds, reason for concern, especially if the CDS tool can link that weight gain to increases in serum glucose readings, suggesting the onset of prediabetes. Granted, a

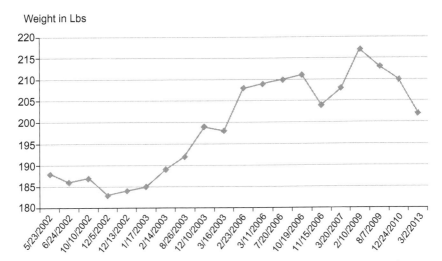

Figure 5.3 Visualizing numerical patient data can help spot worrisome trends more quickly that a list of numbers.

clinician could have detected the same weight issues if the data were a list of numbers, but it would have taken more "cognitive energy."

The somewhat cryptic term "protocol pathway support," feature 4, refers to the use of evidence-based protocols, decision trees, algorithms, and the like. These resources, along with reference materials, are most valuable when clinicians need to slow down. But they are especially useful when they are incorporated into the EHR in a way that makes them easily available. The more mouse clicks needed to reach these resources, the less likely clinicians will use them.

STAR TREK–LIKE DECISION-MAKING TOOLS

It is unlike the average family physician who deals primarily with running noses, enlarged prostates, and broken bones will have much need for some of the more advanced CDS systems, but they certainly have a role to play in medical practice. One of the most promising CDS initiatives that focus on precision medicine is the recently launched Watson Genomics from Quest Diagnostics. More specifically, IBM is bringing Watson's cognitive computing capabilities into the field of oncology in collaboration with Memorial Sloan Kettering Cancer Center, the University of North Carolina Lineberger Comprehensive Cancer Center, and over 20 other cancer institutions. Once a patient's tumor is sequenced to determine its genetic makeup, Watson analyzes this data to look for treatable mutations.

IBM also searches the medical literature, pharmacopeia, and annotated rules created by leading oncologists to look for any relevant research that may suggest a therapeutic approach [35]. The program's ability to scan the mountains of research papers published each year far exceeds any clinician's ability to keep up with the literature.

IBM Watson is not the only group attempting to incorporate genomic data into EHRs and CDSS. The IoM Roundtable on Translating Genomic-Based Research for Health has created a collaborative called DIGITize AC, the goal of which is to accomplish the integration by creating a network of like-minded clinicians, laboratories, vendors, government agencies, standards organizations, and others. Similarly, the National Human Genome Research Institute is involved in initiatives to bring genomic data into EHRs [31,36].

On a less grandiose scale are CDS systems such as VisualDx and Isabel Healthcare. In the past, VisualDx focused on dermatology and visual diagnosis but has now expanded to include cardiopulmonary, gastrointestinal, neurologic, renal, urologic, and infectious diseases. VisualDx can be integrated into many EHRs including Epic using the HL7 Infobutton standard and Cerner using the FHIR standard. By combining its large database of visual images with its Sympticon analyzer, it is capable of deciphering 2700 diagnoses. The Sympticon allows clinicians to compare symptoms with the help of a series of searchable graphics (Fig. 5.4). Research presented at the American Medical Informatics Association Symposium found that VisualDx was capable of doubling the number of correct diagnoses among nondermatologists [37].

The future of CDS will have to include the ability to accommodate all the insights now being collected by the Precision Medicine Initiative, as well as the other precision medicine programs currently in use. A brief review of the list of parameters that likely influence an individual's health status, as outlined in Chapter 1, highlights the difficult task ahead. That list includes the aforementioned genomics and much more, including:

- Diet and physical activity levels
- Alternative therapies being used
- Cardiac and rhythm monitoring
- ICD billing codes
- Problem lists
- Clinical laboratory values
- Imaging studies
- Detailed medication history

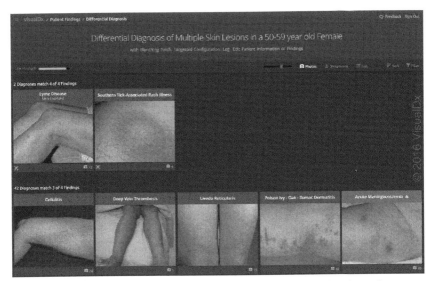

Figure 5.4 Searchable graphics embedded in clinical decision support software can facilitate the differential diagnosis process. *(Reprinted with permission from VisualDx.)*

- Proteomics and metabolites
- Air quality
- Pollutant levels
- Climate changes
- Population density
- Social networking
- OTC medication purchases
- Cell-free DNA
- Infectious exposure

Some of these metrics are already included in currently available EHRs and CDSS, most are not.

REFERENCES

[1] Evans M. For hospitals, a lot of information goes a long way. Wall Str J September 26, 2016:R3.
[2] Cook KA, Modena BD, Simon RA. Improvement in asthma control using a minimally burdensome and proactive smartphone application. J Allergy Clin Immunol Pract 2016;4:730–7.
[3] Kew KM, Cates CJ. Home telemonitoring and remote feedback between clinic visits for asthma (review). Cochrane Database Syst Rev 2016;(8):CD011714.
[4] Shah VN, Garg SK. Managing diabetes in the digital age. Clin Diabetes Endocrinol 2015;1:16.
[5] Pal K, Eastwood SV, Michie S, Farmer A, et al. Computer-based interventions to improve self-management in adults with type 2 diabetes: a systematic review and meta-analysis. Diabetes Care 2014;37:1759–66.

[6] Stratton IR, Adler AI, Neil HW, et al. Association of glycaemia with macrovascular and microvascular complications of type 2 diabetes (UKPDS 35): prospective observational study. BMJ 2000;405:321. http://www.bmj.com/content/321/7258/405.

[7] Data extrapolated from Statton 2000, figure 1.

[8] Cypress M, Tomky D. Using self-monitoring of blood glucose in noninsulin-treated type 2 diabetes. Diabetes Spectr 2013;26(2):102–6.

[9] Klonoff DC. New evidence demonstrates that self-monitoring of blood glucose does not improve outcomes in type 2 diabetes—when this practice is not applied properly. J Diabetes Sci Technol 2008;2(3):342–8.

[10] Shah V HW, Gottlieb P, Beatson C, Snell-Bergeon J, Garg S. Role of mobile technology to improve diabetes care in adults with type 1 diabetes: The Remote-T1d study. Diabetes Technol Ther 2015;17:A25–6.

[11] Quinn CC, Shardell MD, Terrin ML, et al. Cluster-randomized trial of a mobile phone personalized behavioral intervention for blood glucose control. Diabetes Care 2011;34(9):1934–42.

[12] Wild S, Hanley J, Lewis SC, et al. Supported telemonitoring and glycemic control in people with type 2 diabetes: the telescot diabetes pragmatic multicenter randomized controlled trial. PLoS Med July 26, 2016. http://journals.plos.org/plosmedicine/article?id=10.1371/journal.pmed.1002098.

[13] FDA permits marketing of first system of mobile medical apps for continuous glucose monitoring. January 23, 2015. http://www.fda.gov/NewsEvents/Newsroom/PressAnnouncements/ucm431385.htm.

[14] Dexcom. Professional society positioning statements on continuous glucose monitoring system benefits and use recommendations. http://hcp.dexcom.com/sites/dexcom.com/files/styles/pdfs/LBL013821-Rev001-Payer-White-Paper-FINAL-LR.pdf.

[15] Hsu WC, Lau KH, Huang R, et al. Utilization of a cloud-based diabetes management program for insulin initiation and titration enables collaborative decision making between healthcare providers and patients. Diabetes Technol Ther 2016;18:59–67.

[16] Chow CK, Redfern J, Hillis GS, et al. Effect of lifestyle-focused text messaging on risk factor modification in patients with coronary heart disease a randomized clinical trial. JAMA 2015;314:1256–63.

[17] Gotto AM. Jeremiah Metzger lecture: cholesterol, inflammation and atherosclerotic cardiovascular disease: is it all LDL? Trans Am Clin Climatol Assoc 2011;122:256–89. https://www.ncbi.nlm.nih.gov/pmc/articles/PMC3116370/.

[18] American Heart Association. Current science on consumer use of mobile health for cardiovascular disease prevention: a scientific statement from the American Heart Association. Circulation 2015;132:1157–213. https://www.ncbi.nlm.nih.gov/pubmed/26271892.

[19] Khaylis A, Yiaslas T, Bergstrom J, Gore-Felton C. A review of efficacious technology-based weight-loss interventions: five key components. Telemed J E Health 2010;16:931–8.

[20] Hardy ST, Loehr LR, Butler KR, et al. Reducing the blood pressure–related burden of cardiovascular disease: impact of achievable improvements in blood pressure prevention and control. J Am Heart Assoc 2015;4:e002276. http://jaha.ahajournals.org/content/4/10/e002276.full.pdf+html.

[21] Smyth A, O'Donnell M, Lamelas P, et al. Physical activity and anger or emotional upset as triggers of acute myocardial infarction: The INTERHEART study. Circulation 2016;134:1059–67.

[22] Winslow BD, Chadderdon GL, Dechmerowski SJ, et al. Development and clinical evaluation of an mhealth application for stress management. Front Psychiatry 2016;7:130. http://dx.doi.org/10.3389/fpsyt.2016.00130.

[23] Vivify Health. High risk patients. http://www.vivifyhealth.com/childrens-alabama-deploys-vivify-health-solution-closely-monitor-infants-congenital-heart-disease-home/.

[24] Kalantarian H, Motamed B, Alshurafa N, et al. A wearable sensor system for medication adherence prediction. Artif Intell Med 2016;69:43–52.

[25] Marchione FG, Araújob LM, Araújoa LV. Approaches that use software to support the prevention of pressure ulcer: a systematic review. Int J Med Inform 2015;84(10):725–36.

[26] Wang R, Blackburn G, Desai M, et al. Research letter: accuracy of wrist-worn heart rate monitors. JAMA Cardiol October 12, 2016. [Epub ahead of print].

[27] FDA pulse oximeters – premarket notification submissions [510(k)s] guidance for industry and food and drug administration staff. March 2013. http://www.fda.gov/downloads/MedicalDevices/DeviceRegulationandGuidance/GuidanceDocuments/UCM081352.pdf.

[28] Aetna. Pulse oximetry for home use. http://www.aetna.com/cpb/medical/data/300_399/0339.html.

[29] Wang AV, Rogers JR, Jun Y, et al. Use of electronic healthcare records to identify complex patients with atrial fibrillation for targeted intervention. J Am Med Inform Assoc July 3, 2016. pii: ocw082. [Epub ahead of print]. https://www.ncbi.nlm.nih.gov/pubmed/?term=Use+of+electronic+healthcare+records+to+identify+complex+patients+with+atrial+fibrillation+for+targeted+intervention.

[30] Weiss ST. Implementing personalized medicine in the academic health center. J Personal Med 2016;6:18. http://www.mdpi.com/2075-4426/6/3/18.

[31] Aronson SJ, Rehm HL. Building the foundation for genomics in precision medicine. Nature; 2015;536:336–342.

[32] National Academies of Sciences, Engineering, and Medicine. Improving diagnosis in health care. Washington (DC): The National Academies Press; 2015. https://www.nap.edu/catalog/21794/improving-diagnosis-in-health-care.

[33] Graber ML, Franklin N, Gordon R. Diagnostic error in internal medicine. JAMA Intern Med 2005;165:1493–9. http://jamanetwork.com/journals/jamainternalmedicine/fullarticle/486642.

[34] Hummel J. Integrating clinical decision support tools into ambulatory care workflows for improved outcomes and patient safety. September 2013. https://www.healthit.gov/providers-professionals/implementation-resources/integrating-cds-tools-ambulatory-care-workflows.

[35] IBM Watson Health. IBM Watson for genomics. http://www.ibm.com/watson/health/oncology/genomics/.

[36] National Academies of Science Engineering and Medicine. Health and Medicine Division. DIGITizE: displaying and integrating genetic information through the EHR. http://www.nationalacademies.org/hmd/Activities/Research/GenomicBased Research/Innovation-Collaboratives/EHR.aspx.

[37] Papier A, Allen E, Lawson P. Visual informatics: real-time visual decision support. Proc AMIA Symp 2001;987. https://www.ncbi.nlm.nih.gov/pmc/articles/PMC2243669/.

[38] U.S. Food and Drug Administration. Mobile medical applications. http://www.fda.gov/MedicalDevices/DigitalHealth/MobileMedicalApplications/default.htm#b.

[39] U.S. Food and Drug Administration. Examples of mobile apps for which the FDA will exercise enforcement discretion. August 1, 2016. http://www.fda.gov/MedicalDevices/DigitalHealth/MobileMedicalApplications/ucm368744.htm.

i2b2, SHRINE, Clinical Query, and Other Research Tools

Individualizing patient care requires that we be of two minds, one living in the present, the other in the future. We need to think about the best possible care not only for the patient who is currently sitting in the waiting room but also plan for the patients who will need medical services that have yet to be imagined as we analyze data sets and look for new risk factors, diagnostic tools, and treatment options. Clinical research databases, built using software programs such as Informatics for Integrating Biology and the Bedside (i2b2) and the Shared Health Research Information Network (SHRINE), have these twin goals in mind. i2b2 was an NIH-funded National Center for Biomedical Computing based at Partners HealthCare System that developed an informatics framework that allows clinicians and researchers to tap into a large database of deidentified patient records stored in electronic medical records systems. i2b2 was a 10-year grant, which ended a few years ago. It is now funded through a separate entity called the i2b2 Foundation. Also referred to as a software platform or infrastructure, i2b2 consists of a suite of digital tools or modules, including

- A repository of patient records, which holds files of data such as radiological images and genetic sequences.
- A web workbench application and web-based application that let users perform searches.
- An analysis plug-in to detect correlations, which uses mutual information theory to calculate observed correlations within the data of the i2b2 hive.
- Plug-ins to analyze text, export, and import data.
- Software to perform natural language processing, manage projects, and manage ontology [1].
- An identity management cell, the purpose of which is to protect confidential patient information so that any data released to users are consistent with the Health Insurance Portability and Accountability Act of 1996 (HIPAA) privacy rule.

Realizing the Promise of Precision Medicine
ISBN 978-0-12-811635-7
http://dx.doi.org/10.1016/B978-0-12-811635-7.00006-3

Each of these components or modules is linked together into a larger suite referred to as the i2b2 hive, which is illustrated in Fig. 6.1.

The i2b2 software suite is an open-source platform, which means it is freely available to institutions outside of the Partners network. That has enabled over 90 research institutions and academic medical centers and more than 20 international organizations to use the infrastructure as a way to access and search its own electronic health record (EHR) systems and other databases to facilitate clinical research and direct patient care [2].

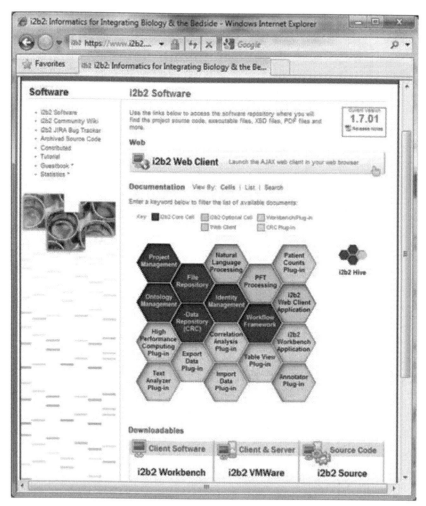

Figure 6.1 The Informatics for Integrating Biology and the Bedside software suite allows Harvard-affiliated hospitals and hospitals around the world to search their patient repositories. *(Permission granted from Shawn Murphy, Partners Healthcare.)*

Partners has also established an i2b2 user community to help expand the value and functionality of the software suite. Shawn Murphy and Adam Wilcox cite the medication extraction challenge as an example of such expansions. It led to improvements in how the software used natural language processing to extract medication information from hospital discharge summaries.

i2b2 has matured and evolved since its inception in 2004. During this time, faculty members have developed a variety of valuable tools to help further the cause of precision medicine. Genephony, for instance, was created to help users analyze large genomic data sets. It allows them to browse genomic data, can be used as an annotation tool, and is a searchable knowledge base. Polymorphism Phenotyping, another component of the i2b2 tool kit, is used to predict the impact of amino acid substitutions on human proteins. When a genetic mutation alters the amino acid structure of enzymes and related proteins, it can have a profound impact on physiology, contributing to disease and its clinical manifestations. As the i2b2 website explains, such predictions are "based on straightforward empirical rules which are applied to the sequence, phylogenetic and structural information characterizing the substitution and can be accessed at http://genetics.bwh. harvrd.edu/pph/."

TAKING INFORMATICS FOR INTEGRATING BIOLOGY AND THE BEDSIDE INTO NEW TERRITORY

Making i2b2 freely available to other institutions has generated unique and imaginative "offspring" that the original developers of the platform never anticipated. One unexpected offspring relates to the design of cancer trials.

One of the major problems in developing and launching the clinical trials that fuel both personalized and population-based patient care is the inability to recruit enough patients to reach statistically significant results. The Institute of Medicine has estimated that 40% of cancer cooperative group trials never get off the ground for this reason [3]. The waste incurred by such failed efforts is not trivial. When researchers attempt to locate enough patients to meet statistical requirements, as well as the various inclusion and exclusion criteria of the study protocol, they often keep the subject accrual process open for many months, which translates into expenses on personnel to setup and operate the initiative. In today's health-care environment, cost savings have to be taken into account at every turn.

To help solve the patient recruitment problem at the Kimmel Cancer Center of Thomas Jefferson University in Philadelphia, London et al. decided to make use of the University's clinical data warehouse and research data mart (RDM), which uses the i2b2 framework. Essentially, they attempted to predict whether their proposed clinical trials would be able to find enough patients by looking at the data already in the RDM of deidentified patient records.

Jefferson's data mart included an invaluable collection of facts and figures about patients who had already been treated at the cancer center. It included 28 million observations on approximately 350,000 patents and over 800,000 biological specimens. It also contained patient demographics, diagnoses, the procedures they underwent, their lab results, medications, hospital stays, and vital signs. Data on the biological specimens told investigators where tumors were located, whether they were fluid or solid in nature, and each patient's cancer recurrence, survival, and treatment choices.

The researchers collected the eligibility criteria for 90 cancer trials. These same criteria were also fed into the i2b2-based data mart to see if this tool could help determine which of the cancer trials had a reasonable chance of attracting the necessary patients for the studies who would meet the criteria. Using this methodology, Jefferson's data mart was able to generate a positive predictive value of 95.8%, which meant that 23 of the 24 trials that the database had predicted would not locate enough patient recruits actually did not locate those patients. Put another way "if the methodology predicts 'failed accrual,' then we should trust this prediction and should not proceed to open the trial with its current eligibility criteria," according to London et al. On a less positive note, if the database analysis suggested that a new clinical trial would successfully locate all its needed patients, there was little guarantee that the target numbers would be reached since the sensitivity of the methodology was only 39.7%.

i2b2 has also been of more direct value in clinical research, helping investigators explore the relationship between adverse drug effects and various disease states. Oftentimes, such relationships are revealed long before the FDA issues a black box warning for a drug or removes the drug from the market. For example, the agency issued a black box warning about the antidepressant citalopram (Celexa) in 2011, informing clinicians and the public that the drug could cause a prolonged QT interval on an EKG, which has been associated with arrhythmias, fainting, seizures, and sudden death. Clinicians suspected the problem much earlier.

When clinicians see such warnings, they frequently switch to other antidepressants without such black box labeling. Many switched to escitalopram (Lexapro). However, a data analysis that took advantage of the patient records stored in i2b2 concluded that escitalopram also caused prolonged QT interval syndrome. Victor Castro, with Partners Research Computing, and his associates [4] analyzed EHRs from patients taking citalopram, escitalopram, and amitriptyline. They were able to review nearly 1.5 million QTc measurements from EKG tracings. Among the more than 38,000 patients who had EKGs performed after receiving antidepressants, Castro et al. found a dose-response association between QTc prolongation and citalopram, escitalopram, and amitriptyline, suggesting that the risk of life-threatening arrhythmias are as much of a concern for escitalopram, for which no black box warning about long QTc existed, as they are for citalopram, for which there was.

SHARED HEALTH RESEARCH INFORMATICS NETWORK EXTENDS THE INFORMATICS FOR INTEGRATING BIOLOGY AND THE BEDSIDE REACH

The hospitals affiliated with Harvard Medical School, which include Massachusetts General Hospital, Brigham and Women's Hospital, Beth Israel Deaconess Medical Center, Children's Hospital Boston, and Dana Farber Cancer Center, each has i2b2 installed on their computers, and each hospital has been able to use the software suite to search its respective EHR system to inform patient care and generate hypotheses for clinical research. In 2009, a decision was made to link all the medical records from all five hospitals so that clinicians and researchers would have access to a much larger source of patient data. SHRINE, this combined source, is a federated web-based query system that allows qualified members of the Harvard community to cast a much wider net as it seeks to answer to puzzling therapeutic and diagnostic questions and as it considers the formulation of new clinical research projects [5]. SHRINE is available to members of the five hospitals and faculty at Harvard Medical School. However, since the SHRINE software is freely available as open source, it can be used by other medical centers to establish similar data sharing networks.

The SHRINE website outlines some of the possible uses for the network:
- Generating new research hypotheses
- Planning research requiring large sample sizes not easily available at any single institution

- Preparing grant applications that would benefit from preidentification and/or characterization of a potential research cohort
- Identifying potential cohorts for clinical trials
- Conducting research in the areas of population health and health services

Kenneth Mandl, MD, MPH, an Associate Professor of Pediatrics at Boston Children's Hospital, discusses the other function that is the most relevant to our discussion of precision medicine: "I use SHRINE to investigate personalized therapies for patients. Rather than relying on clinical trials data as a source of evidence, the approach is to examine the real-world experience of patients similar to ours. This is a shift toward using large-scale observational data sets to form the evidence base." [6].

Nonetheless the value of SHRINE in creating patient cohorts large enough to justify clinical trials and observational studies is probably the most impressive feature of the network. For example, in a review of the functionality of SHRINE, Andrew McMurry and his colleagues [7] use the illustration of a researcher at Boston Children's Hospital who wants to analyze the effectiveness of chemotherapeutic drugs in patients with acute lymphoid leukemia. Since the cancer is so rate, it would be difficult to rely solely on data from her hospital to do the analysis or prepare for a clinical trial. Instead, she can request a query approval from the SHRINE local data steward. Once approved, she can input the query into the online dashboard and obtain an aggregated patient set, tapping into the records of the five Harvard-affiliated hospitals that participate in SHRINE. She can further refine her search by requesting all patients who were treated with a multiple drug regimen and who had complete blood counts to confirm their diagnoses. Fig. 6.2 shows a screen capture of said SHRINE query results, which yields about 1134 patients that fit all three criteria.

It is important to note in Fig. 6.2 that the aggregated patient total is 1134 ± 12. That range is the result of a build-in mechanism that protects patient privacy. The precise number of patients in any SHRINE query result is shrouded and the counts given to each user are only an estimate, based on an algorithm published by Partners health-care researchers [8]. The obfuscation algorithm adds a small random number to the actual count generated by the SHRINE query.

SHRINE and Clinical Query never report the exact number to ensure that the privacy of individual patients is protected because it is possible for one to create a query so arcane that it identifies a single individual. For example, consider this fictional query: "My neighbor has one blue and one green eye.

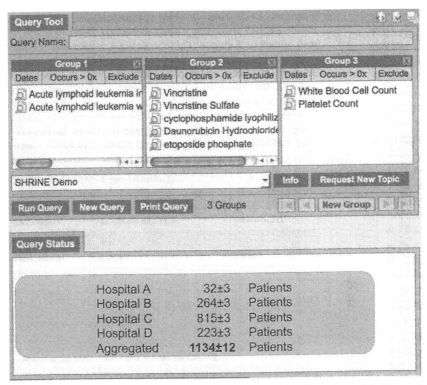

Figure 6.2 An example of search results performed in Shared Health Research Information Network. *(From McMurry AJ, Murphy SN, MacFadden, D et al. SHRINE: enabling nationally scalable multi-site disease studies. PLoS One March 2013;8(3):e55811.)*

Show me the count of all blue-eyed, green-eyed people taking mental health medications." A count of one could reveal my neighbor's identity and the fact that he uses said medications. By adding arbitrary numbers to every result, the tool ensures that population queries remain ambiguous. That makes it virtually impossible for a user to guess the identity of an individual patient based on the demographic and related details in the deidentified record. Additional precautions taken to protect patient privacy include encryption using World Wide Web Consortium standards (W3C), removal of all personally identifiable details, and the fact that SHRINE queries only return aggregate counts.

THE IMPACT OF SHARED HEALTH RESEARCH INFORMATION NETWORK ON SCIENTIFIC RESEARCH

SHRINE and i2b2 have both proven to be valuable informatics tools to help researchers solve puzzles that were unsolvable using smaller databases. For instance, they allowed Issac Kohane and his associates at Harvard's Center for Biomedical Informatics to derive more accurate statistics on the incidence of comorbidities among children and young adults with autism [9].

Their retrospective prevalence study used EHR data on more than 14,000 individuals with autism spectrum disorders who were cared for at three general hospitals and one pediatric hospital in Boston. The entire population from which these cases were derived consisted of more than 2.3 million patients below age 35 years. Most previous investigations performed to determine what coexisting disorders were common in autistic patients consisted of fewer than 200 individuals, making it almost impossible to get an accurate estimate and increasing the likelihood of a type 2 statistical error. (And as we mentioned elsewhere in this book, gathering data on a very large patient population allows clinicians to identify *subpopulations* with more certainty, subpopulations that may have unique risk factors and require different—and more personalized—treatment protocols).

Data networks such as SHRINE also help clinicians and health-care organizations entering the world of pay-for-value, in which they are expected to be more accountable and to take on more financial risk. Without a more detailed understanding of the coexisting disorders that autistic patients are likely to experience, it is much more difficult to plan their care and to factor the cost of that care into the risk equation.

Kohane et al. found that patients with autistic spectrum disorders are far more likely to also have epilepsy (11.4% vs. 2.2%). Similarly the prevalence of schizophrenia among autistic patients was 2.4%, compared to only 0.2% in the overall hospital population. The prevalence of bowel disorders, excluding inflammatory bowel disease, was 11.7% versus 4.5% in controls. Although the size of the SHRINE data set helped generate more reliable statistics on comorbidities, the researchers did acknowledge that the source of the data, namely (International Classification of Diseases) ICD-9 codes, has its limitations, stating that "This study uses ICD-9 codes (the controlled vocabulary employed principally by healthcare providers to bill for their services), of which some are clearly disease states while others are codes representing symptom complexes. Without chart review we cannot determine if, for example, a diagnosis is determined by symptoms or by diagnostic tests. The ICD-9 codes are also coarse grained and because there are often used for billing, they therefore have potential for bias."

SHRINE has also been used to help monitor the incidence of posttreatment complications among patients with multiple myeloma. Jeremy Warner, from the Department of Medicine, Division of Hematology and Oncology, Vanderbilt University, Nashville, and associates [10] initially analyzed patient records from the Multiparameter Intelligent Monitoring in Intensive Care (MIMIC II) database, which includes hospital ICU admissions that took place between 2001 and 2007, to compile a list of complications associated with multiple myeloma. They also ran a SHRINE query for all myeloma patients based on the 8 ICD-10 codes related to the cancer. Using data from both sources, Warner et al. were able to determine that the rate of noninfectious treatment complications had increased from 2001 to 2007 by 6%. This proof of principle study demonstrated that SHRINE can play a useful role in answering specific questions that are relevant to health services research.

BETH ISRAEL DEACONESS MEDICINE AND CLINICAL QUERY 2

Clinical Query 2, illustrated in Fig. 6.3, is one of the highlights of Beth Israel Deaconess Medical Center's medical informatics platform. Like SHRINE, CQ2 is based on the i2b2 software suite. Each Harvard-affiliated hospital has a local query tool for its researchers. CQ2 is the name that BIDMC has given its local i2b2 implementation. CQ2 only accesses Beth Israel's patient population. The data in SHRINE start in 2001 and the data are always 6 to 12 months old. CQ2's data, on the other hand, start in 1997 and they are refreshed every 0–2 months. The depth of CQ2 also contributes to its value. SHRINE contains demographics, diagnoses, medications, and some lab tests. CQ2 includes all lab tests, procedures, and vital signs. In fact, the local systems of every hospital that are part of Harvard SHRINE have more data and are updated more often than Harvard SHRINE. Despite these advantages, the differences are getting narrower over time. Both SHRINE and CQ2 are based on the open-source software platform i2b2 and in the past, SHRINE used a much older version of i2b2 than CQ2. However, SHRINE is slowly catching up.

It should also be pointed out that most institutions that participate in the federated SHRINE/ Patient-Centered Clinical Research Network (PCORNet) networks also have the capability of running queries on local systems. The data and functionality of the local versus federated systems often differ, and investigators need to choose which one is most appropriate for their studies.

To use SHRINE, one needs to first go through an approval process. In addition, fellows and postdoctoral researchers must obtain a faculty sponsor

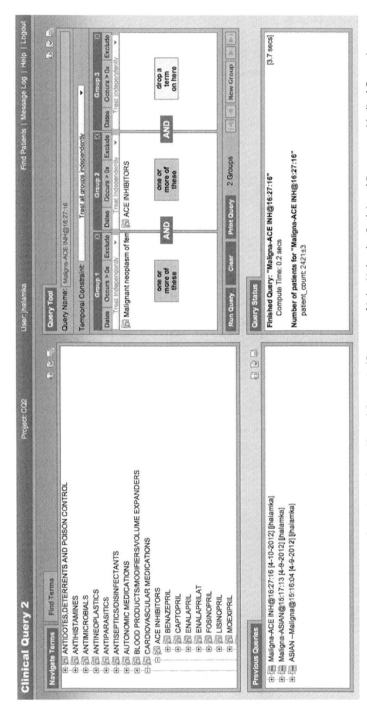

Figure 6.3 An example of search results in Clinical Query 2. *(Courtesy of John Halamka, Beth Israel Deaconess Medical Center.)*

to use it. CQ2 can be used by anyone at BIDMC who has HIPAA training. And finally, CQ2 allows users to extract patient level data sets, while SHRINE only provides aggregate counts.

MOVING TOWARD A NATIONAL NETWORK

Clinical Query serves the needs of clinicians and researchers at BIDMC; SHRINE serves the needs of all the Harvard-affiliated hospitals, and i2b2 serves as the backbone or infrastructure for both digital tools. But the need for medical informatics and the contribution it can make to population-based and personalized medicine warrant a much larger palette. The National PCORnet, which is part of the Patient-Centered Outcomes Research Institute (PCORI), has been established to provide much broader brush strokes.

The Institute was created to address the many unmet needs of patients, needs that have not been adequately met by traditional medical research. To meet the challenge, PCORI is attempting to "close the gaps in evidence needed to improve key health outcomes. To do this, we identify critical research questions, fund patient-centered comparative clinical effectiveness research, or CER, and disseminate the results in ways that the end-users of our work will find useful and valuable." The Institute has established several programs to accomplish these goals, one of which is a research infrastructure initiative, which has resulted in the creation of the PCORnet.

Like SHRINE and Clinical Query, PCORnet aims to find faster, less expensive, and more powerful ways to conduct comparative effectiveness research. And one way it hopes to do that is by harnessing the untapped power of EHRs through a network of networks that allows researchers to conduct large observational studies and large pragmatic clinical trials. In addition, PCORnet is tapping patient-generated information and insurance claims data.

This initiative consists of 13 clinical data research networks and 20 patient-powered research networks. By tapping the resources of all these diverse groups, PCORnet is gleaning practical insights based on the experiences and clinical records of more than 145 million people! [11] It is easy to quickly read these statistics without considering their weight. The current US population is about 325 million, which means the PCORnet database includes nearly *half* of all Americans (44%).

The clinical data research networks include groups sponsored by the University of Kansas Medical Center, Kaiser Foundation Research Institute, Vanderbilt University, The Children's Hospital of Philadelphia, Harvard

University, Mayo Clinic, and several others. The patient-powered research networks include initiatives sponsored by the Epilepsy Foundation, the Global Health Living Foundation, COPD Foundation, Massachusetts General Hospital, Mayo Clinic, the Genetic Alliance, and several others.

As of December 2016, PCORI has awarded $1.61 billion to fund 570 comparative effectiveness studies and related projects. The Institute refers to these initiatives as "Research Done Differently." By way of contrast to many traditional medical research initiatives, these projects include a focus on top- ics, questions, and outcomes that *patients* really care about, not just research- ers. These projects also get patients more directly involved in the research process, not just as subjects but as partners and planners. The approach gives a whole new meaning to the term patient engagement [12].

Of course, to make this approach a reality requires more than a new philosophy of science, it demands creative ways of handling medical data. With that in mind, the network has created the PCORnet Common Data Model. Essentially it is a way to allow data from a diverse group of health- care systems, many using different computer languages, to communicate with PCORnet so that researchers using the PCORnet data will be reading data written in one common language.

If, for example, the Children's Hospital of Philadelphia codes a common term like a patient's first name as "F_Name" and the Genetic Alliance uses "First_Name," it becomes difficult for a researcher to use this data. But the Common Data Model (CDM) requires all the research and patient net- works that want to contribute to a PCORnet investigation to "map" or link their data to the PCORnet database by using the CDM standards. Among the standards that constitute the CDM are widely accepted terminologies, including ICD, (Systematized Nomenclature of Medicine) SNOMED, (Current Procedural Terminology) CPT, and (Logical Observation Identifiers Names and Codes) LOINC.

Unfortunately, bringing together 33 clinical research and patient- oriented networks together is like bringing delegates from around the world to the United Nations. Without a robust group of translators, we are faced with a modern-day Tower of Babel. The PCORnet Common Data Model is somewhat equivalent to those UN translators.

More specifically, the CPORnet CDM is based on the FDA's Sentinel Initiative Common Data Model. Sentinel is a surveillance system sponsored by the federal agency to help it monitor the safety of medical products. As is the case with CPORnet, the FDA has to communicate with a wide network of partners that use various computer languages. It created its own data model to prevent its own Tower of Babel.

PATIENT-CENTERED OUTCOMES RESEARCH INITIATIVES

PCORI has launched several research projects that demonstrate the benefit of its collaborative, EHR-enabled approach. The first study, called Aspirin Dosing: A Patient-Centric Trial Assessing Benefits and Long-Term Effectiveness (ADAPTABLE), is enrolling as many as 20,000 patients with heart disease to determine the optimal dose for preventing myocardial infarction and stroke. This comparative effective trial compares low-dose 81 mg of aspirin to the standard 325 mg dose. Patients at risk for either condition are being assigned to one of the two doses and followed for 30 months, with investigators at Duke University taking the lead on the project. Six clinical data research networks and one patient-powered research network are being tapped to contribute data to the research.

ADAPTABLE is taking a precision medicine approach and plans to avoid the usual research findings, which offer advice for the average patient taking aspirin. The Duke investigators are looking at benefits and adverse effects parsed by gender, age, ethnicity, and by comorbidities such as diabetes. With a patient sample approaching 20,000, this type of subgroup analysis should be possible without compromising statistical integrity.

The ADAPTABLE study is the first pragmatic clinical trial launched by PCORI. It uses patients' digital records and web-based technology to allow participants to self-consent and report data directly into an online portal.

As we have said in previous chapters, patient engagement is a key component of personalized medicine. Without buy-in from individual patients, any attempt to personalize care will be thwarted at every turn. One of the reasons patients do not engage in their own care is they do not sense the patient/doctor relationship is a partnership among equals. Despite efforts to "depaternalize" patient care in recent years, some clinicians continue to view their patients as children who should simply follow orders. In this type of environment, many patients fear that disagreeing with their provider will result into retribution in the form of inferior care. One of the new PCORI-funded studies is addressing this fear head-on. Ming Tai-Seale, PhD, MPH, is the lead investigator on this 36-month project, which has been spearheaded by the Palo Alto Medical Foundation Institute. The project involves the development and testing of materials that teach physicians and patients to work together to reach decisions.

The training has three components: A video to help patients prepare for a doctor's visit, which drives home the message that it is acceptable to ask questions, to occasionally interrupt the clinician, and to ask about alternative treatment options. The program also provides patients with a booklet that they can take with them when they get into the office, in which they write down things they want to talk about. Thirdly, physicians participating in the study have conversations with actors portraying patients who disagree with them about their recommendations.

Three hundred patients saw their primary care practitioners with and without receiving the training before their office visits and they recorded their impressions on surveys. While the final study results were not available as we went to press, Tai-Seale says "Our patient-centered intervention resulted in better patient experience with care and with shared decision making." [13].

PCORnet has also funded obesity studies that address its prevention and treatment. One investigation is looking at bariatric surgery, comparing three procedures to assess their benefits and adverse effects. A second project will try to determine if various types of antibiotics, when administered at a very early age, have any impact on weight gain as children mature [14]. These observational studies are also taking advantage of the large database the Institute has been able to create based on EHR data from numerous organizations.

The bariatric surgery study will examine the advantages and disadvantages of adjustable gastric band, roux-en-gastric bypass, and vertical sleeve gastrectomy and plans to analyze the records of 60,000 patients who have had one of these procedures. It will include 17,000 diabetics and 900 teens. The antibiotic study will analyze the records of approximately 600,000 patients to see if various types of antibiotics affect children's growth and weight at ages 5 and 10 years.

PCORnet's parent organization, PCORI, has also been funding a wide variety of projects that put more emphasis on precision medicine. For example, David M. Kent, MD, MS, of Tufts Medical Center, is the principal investigator for a project entitled: How Well Do Clinical Prediction Models (CPMs) Validate? A Large-Scale Evaluation of Cardiovascular CPMs. In earlier chapters, we have discussed the value of data-driven predictive tools to help personalize medical care. Kent's investigation takes a closer look at the role of such clinical predictive models to determine how effective they are in pinpointing an individual's risk of disease in the real world, and how often these tools are actually being used in clinical practice. The PCORI summary of the project emphasizes the need to critically examine the role of these predictive models [15].

Surprisingly little is known about how accurately CPMs generally tend to predict, with respect to their intended use. Most CPMs are not rigorously tested; when they are, they are often tested by the same team that developed the CPM, and reporting can be incomplete or biased. For example, while it is often reported how well a CPM can separate patients with and without the outcome of interest (a measure referred to as "discrimination"), far less frequently is it shown how close the predicted risks are to the actual outcome rates across relevant groups of patients (a measure referred to as "calibration"). Using "miscalibrated" CPMs can lead to misinformation and harmful decisions. …

Our long-term objective is to motivate a culture of routine, rigorous, transparent, and fully independent CPM testing and continual updating to support accurate prediction for patient-centered decision making. Our short-term objectives are to describe the current state of evaluation studies of cardiovascular CPMs, to understand how well CPMs perform on independent datasets generally, and to understand the effectiveness of updating procedures across a broad array of CPMs. We also seek to gain insight into when CPMs are likely or unlikely to be robustly transportable to different settings. Finally, our project will create a comprehensive online open resource that provides stakeholders with information on which cardiovascular CPMs appear to provide the most reliable and robust predictions.

REFERENCES

[1] Informatics for Integrating Biology & the Bedside. I2b2 software. 2016. https://www.i2b2.org/software/index.html.

[2] Murphy S, Wilcox A. Mission and sustainability of informatics for integrating biology and the bedside (i2b2). eGEMS September 11, 2014;2(2):1074. https://www.ncbi.nlm.nih.gov/pubmed/25848608.

[3] London JW, Balestrucci L, Chatterjee D, et al. Design-phase prediction of potential cancer clinical trial accrual success using a research data mart. J Am Med Inform Assoc 2013;20:e260–6.

[4] Castro VM, Clements CC, Murphy SN, et al. QT interval and antidepressant use: a cross sectional study of electronic health records. BMJ 2013;346:f288.

[5] Harvard Catalyst. SHRINE: at a glance. https://catalyst.harvard.edu/services/shrine/.

[6] Harvard Catalyst. Spotlights: SHRINE. http://catalyst.harvard.edu/spotlights/shrine.html.

[7] McMurry AJ, Murphy SN, MacFadden D, et al. SHRINE: enabling nationally scalable multi-site disease studies. PLoS One March 2013;8(3):e55811.

[8] Murphy SN, Chueh HC. A security architecture for query tools used to access large biomedical databases. AMIA Annu Symp Proc 2002. https://catalyst.harvard.edu/docs/SHRINE/procamiasymp00001-0593.pdf.

[9] Kohane IS, McMurry A, Weber G, et al. The co-morbidity burden of children and young adults with autism spectrum disorders. PLoS One 2012;7(4):e33224.

[10] Warner JL, Alterovitz G, Bodio K, et al. External phenome analysis enables a rational federated query strategy to detect changing rates of treatment-related complications associated with multiple myeloma. J Am Med Inform Assoc 2013;20(4):696–9. https://www.ncbi.nlm.nih.gov/pubmed/?term=External+phenome+analysis+enables+a+rational+federated+query+strategy+to+detect+changing+rates+of+treatment-related+complications+associated+with+multiple+myeloma.

[11] PCORI. PCORnet: the national patient-centered clinical research network factsheet. http://www.pcori.org/sites/default/files/PCORI-PCORnet-Fact-Sheet.pdf.

[12] PCORI. Research done differently fact sheet. http://www.pcori.org/sites/default/files/PCORI-Research-Done-Differently.pdf.

[13] PCORI. Creating a "zone of openness" at the doctor's office. http://www.pcori.org/research-in-action/creating-zone-openness-doctors-office.

[14] Patient Centered Outcomes Research Institute. Fact sheet: PCORnet's first observational studies: addressing questions on the treatment and prevention of obesity. http://pcornet.org/wp-content/uploads/2015/08/Fact-Sheet-PCORI-Obesity-Studies.pdf.

[15] PCORI. How well do clinical prediction models (CPMs) validate? A large-scale evaluation of cardiovascular clinical prediction models. http://www.pcori.org/research-results/2016/how-well-do-clinical-prediction-models-cpms-validate-large-scale-evaluation.

CHAPTER SEVEN

Barriers and Limitations

There is certainly no paucity of critics who maintain that the quest to achieve personalized medicine is a mistake, or unattainable, or both. The criticisms often revolve around the issues of cost, the additional workload for clinicians, lack of adequate scientific evidence to support its efficacy, lack of genomics expertise among primary care providers, and many other concerns. Our purpose here is to address these concerns.

DOES PRECISION MEDICINE COST TOO MUCH?

In Chapter 3, we discussed the role of pharmacogenomics and bio-markers in precision medicine. The FDA has approved a long list of medications that have been associated with genetic variants and for which specific biomarkers have been identified [1]. Testing positive for specific biomarkers is one way in which clinicians are able to personalize the treatment of cancer, heart disease, and other disorders. These biomarkers can help practitioners determine genotype-specific dosing, estimate the risk of adverse effects, and approximate a patient's therapeutic response or lack thereof.

But the problem with many of these drugs is their high cost, and the fact that the cost-benefit ratio is not always very positive. Imatinib (IM) (Gleevec), which is indicated for chronic myeloid leukemia (CML) and trastuzumab (Herceptin), which is indicated for the treatment for HER2-positive breast cancer and certain gastric and gastroepophageal cancers, have proven cost-effective. Other agents have not [2].

Hannah Bower and her colleagues at the Karolinska Institute in Sweden analyzed data on IM among patients in the Swedish Cancer Registry from 1973 to 2013 and found the drug was responsible for dramatic improvements in life expectancy. For example, a 55-year-old male CML patient was expected to live 3.5 years in 1980, before the drug was introduced into clinical practice (In the 1980s, busulfan and hydroxyurea were used to treat CML; that was followed by stem cell transplantation and interferon alfa in the 1980s and 1990s. IM was introduced around 2000). Life expectancy of

Realizing the Promise of Precision Medicine
ISBN 978-0-12-811635-7
http://dx.doi.org/10.1016/B978-0-12-811635-7.00007-5

the same 55-year-old male was an additional 27.3 years by 2010! Bower et al. summarize their findings among all age groups with the cancer by saying "The life expectancy of CML patients of all ages diagnosed in 2010 was within 3 years of the life expectancy of the general population…" [3]. And while their investigation was not designed as a cost analysis, Ohm et al. reviewed the cost-effectiveness ratios of IM in CML and concluded "that incremental cost-effectiveness ratios that compared IM with other treatments were generally acceptable by health authorities, which means that these treatments should continue to be financially feasible" [4].

It is also worth discussing the mechanism of action of IM since it is relevant to our premise regarding the difference between precision medicine and one-size-fits-all medicine. In two illustrations in Chapter 1, we compared to management of rheumatoid arthritis with tumor necrosis factor alpha inhibitors to the management of simple iron deficiency with iron supplements and diet. The former only interrupted the disease at a midpoint in its pathophysiology, while the latter addressed the underlying root cause of the disorder. IM comes close to addressing the underlying cause of CML, making it a more precise treatment option. In CML, there is an acquired balanced chromosomal translocation[1] that results in an active tyrosine kinase called BCR-ABLI. IM mesylate targets the BCR-ABLI oncoprotein.

Similarly, research on trastuzumab and related anti-HER2 monoclonal antibodies has demonstrated the profound benefits of these precision cancer drugs. Sandra Swain, MD, from the Washington Cancer Center in Washington DC, and her colleagues, have shown that a combination of pertuzumab, trastuzumab, and docetaxel have significant benefits in women with HER2-positive metastatic breast cancer [5]. They found that the median overall survival of women taking pertuzumab—also a HER2 monoclonal antibody—trastuzumab, and docetaxel was 56.5 months, compared to only 40.8 months in women in the control group, who were taking placebo, trastuzumab, and docetaxel. Previous studies that included trastuzumab plus chemotherapy resulted in a median overall survival of 25.1–38.1 months.

Unfortunately, these success stories have not been duplicated across the board. For example, there were approximately 70 new approvals for solid tumors issued by FDA between 2002 and 2014 [6]. The therapeutic value of

---

[1] During a chromosomal translocation, part of a chromosome moves from its normal location to another location on a second chromosome. Since the genes located on chromosomes are responsible for the synthesis of body proteins, including enzymes, the resulting change in DNA structure can cause havoc with the formation of the enzymes needed to carry out normal body functions.

these cancer drugs is less than impressive. They resulted in a median increase in progression-free survival of 2.5 months. Similarly, overall survival only increased by 2.1 months. Of course, deciding if 2–2.5 months of additional life is worthwhile is a highly personal matter, but when one adds in the out-of-pocket costs many patients must now contend with as a result of the surge in high deductible insurance plans, and then consider the discomfort patients must cope with from the disease and its treatment, one must question whether these agents have improved quality of life for patients and their families. Weighing all these variables, it is evident that determining the value of personalized medicine is itself a personal issue in this context, depending on each patient's perspective.

Nonetheless, working groups convened by the American Society of Clinical Oncology do not consider an additional 2 months of survival clinically meaningful. When asked to offer their suggestions on how to design clinical trials that would significantly improve survival, quality of life, or both, the experts concluded that for pancreatic cancer, they would want to see at least a 4 month improvement in overall survival (The current overall survival is 10–11 months). For lung cancer, they expected at least an additional 3.25 months for nonsquamous cell carcinoma (Current survival is 13 months). For metastatic triple negative breast cancer, at least an additional 4.5 months (Current survival is 18 months). Using these very modest gains as a guideline, only 30 of the 71 FDA approvals mentioned above would offer clinically meaningful improvements [6]. Fojo et al. state: "…this highlights the fact that 1 important reason for the rise in the number of FDA approvals has been a lowering of the efficacy bar—a lowering that, while providing a greater number of options for patients, has resulted in the approval of therapies that many would argue did not achieve clinically meaningful improvement." They go on to conclude that "the eventual revenue from such therapies rather than their need has been a driving force cannot be doubted."

Other economic reviews of precision cancer medications have rendered similar conclusions. An analysis in the *Journal of Economic Perspectives* [7] that reviewed the cost and benefits of 58 oncology drugs approved by the FDA from 1995 to 2013 concluded: "The scientific knowledge embodied by new drugs is impressive, but progress in basic science has not always been accompanied by proportionate improvements in patient outcomes. Gains in survival time associated with recently approved anticancer drugs are typically measured in months, not years." What is even less encouraging is the fact that it is getting more expensive to pay for each additional month of life that these medications are providing. Insurance companies and patients paid $54,100 for a year of life in 1995, but in 2005, it costs $139,000. In 2013, the same year of life cost $207,000.

Similar concerns have been raised with regard to drugs recently approved to more precisely target cystic fibrosis (CF). The disease, which affects about 1 in 3500 live births, has proven decidedly difficult to manage, with most treatment protocols targeted at the signs and symptoms rather than underlying causes. But the FDA has approved ivacaftor, which seeks to correct the CF transmembrane conductance regulator dysfunction. It has been shown to inhibit the pulmonary complications of the disease, normalize sweat chloride levels, and improve lung function and quality of life [8]. But these benefits have only been documented in patients with the specific mutation that the drug addresses, namely a G551D mutation, which exists in only about 4% of CF patients. Even more troubling is the cost: More than $300,000 per patient per year.

FDA has also approved a combination drug called Orkambi, which includes both ivacaftor and lumacaftor; it targets a far more common mutation seen in CF patients. However, according to Ferkol and Quinton, "This combination... produced a modest clinical improvement in the primary endpoint, FEV1 [forced expiratory volume], in adults and adolescents with CF homozygous for the delF508 mutation. This drug did not appear to be better than conventional therapies." The new drug costs $258,000 per patient per year, or as much as $15 million over a CF patient's life.

There remains one weakness in such cost critiques: Precision drug therapy that targets genetic flaws is not precision medicine, it is only one component of it. As we have discussed in previous chapters, a truly personalized approach to patient care involves a holistic perspective that looks at all the parameters that influence one's disorder, including one's genome, exposure to environmental toxins, stress levels, nutritional status, and so on. This is also the approach that the federally sponsored Precision Medicine Initiative is taking as it gathers data on a wide array of genetic and environmental risk factors.

Additionally, there is some research to suggest that precision drug therapy may in fact be cost-effective. Haslem et al. compared 36 patients with a variety of solid tumors who received precision cancer therapy to 36 historic controls, which included patients receiving standard chemotherapy or supportive care. The patients receiving precision medicine therapy had recurrent/metastatic tumors and had failed earlier treatment protocols. They were all cared for through the Intermountain Healthcare delivery system and received genomic testing to detect mutations that would respond to specific antitumor agents known to address the genetic defects [9].

Average progression-free survival was 22.9 weeks among patients in the precision medicine group, compared to 12 weeks in the control group ($P = .002$), which translates into about three additional months of life. That difference is certainly not impressive, but it came at a more reasonable cost. The charges per patient per week were $4665 in the precision medicine cohort versus $5000 in the control group.

There is also evidence to suggest that pharmacogenetic profiling results in cost-effective patient care. In a 2017 randomized prospective study, pharmacists applied their knowledge of drug/gene-based interactions to a group of patients aged 50 and older who were each taking numerous medications and managed at home after hospital discharge, comparing them prospectively to a control group who received standard drug information. The relative risk of 30-day hospital readmission for patients whose medication was analyzed to detect variations in the cytochrome P450 genes responsible for the enzymes that metabolize most commonly prescribed drugs were 0.65, compared to the control group. That improved outcome continued at 60 days with a relative risk of 0.48. The mean number of ED visits per patient was 0.25 for genetically profiled patients compared to 0.40 for the controls (RR 0.62). Although actual cost and charge data were not available for this investigation, Elliott et al. used modeling to estimate costs based on Medicare average all-cause readmission and ED visit costs. That generated a $4382 per patient savings at 60 days [10]. More details on the cost analysis are available in Table 7.1.

Table 7.1 Estimated financial savings from rehospitalizations and emergency department visits reduction

Outcomes	Mean number of events per patient	Average cost per event	Average cost per patient	Per patient savings vs. untested
Untested: rehospitalization	0.70	$11,200	$7,840	
Tested: rehospitalization	0.33	$11,200	$3,696	$4,144
Untested: ED visits	0.66	$884	$543	
Tested: ED visits	0.39	$884	$345	$238
Total per patient savings in 60 days prior to cost of intervention				$4,382

Medicare average all-cause readmission cost in 2009 was $11,200 and average ER visit cost in 2011 was $884. Model based on Medicare average showed a $4382 per patient cost savings in 60 days balanced by a test cost of $914. *ED*, Emergency Department; *ER*, Emergency Room.
From Elliott LS, Henderson JC, Neredilek MB, et al. Clinical impact of pharmacogenetic profiling with a clinical decision support tool in polypharmacy home health patients: a prospective pilot randomized controlled trial. PLoS One 2017;12(2):e170905.

FILLING THE GENOMICS INFORMATION GAP

Several thought leaders have pointed out that full adaption of personalized medicine in clinical practice requires a more in-depth understanding of genetics and genomics than the average physician possesses. For instance, a 2012 study that queried more than 2400 primary care physicians (PCPs) in North Carolina received 382 responses; among these respondents, only 38.7% said they were even aware of direct-to-consumer genetic testing; among those who were, only 15% felt capable of discussing it with patients [11]. Similarly a survey that asked European PCPs about their confidence in discussing basic medical genetic tests revealed only 19.3% were confident or very confident in their ability to discuss them.

While it is still possible for patients who need genetic testing or counseling about test results to turn to specialists in the field, there too we run into problems. In 2015, about 3000 genetic counselors were employed in the United States, and 135 were practicing outside the country [11]. Statistics from the American Board of Medical Genetics and Genomics indicate that as of July 2014 there were only 1286 certified clinical geneticists in the world, 1194 of which practiced in the United States. Putting that figure into context that is 0.18% of the total number of physicians in the nation.

Although these statistics are discouraging, they do not justify slowing down the advance of precision medicine. It is not as though every patient who walks into a PCP's office will actually need a genetic analysis. To date, there are still too few clear cut associations between gene variants and specific diseases, with the fields of oncology and pharmacogenomics being two exceptions. Nonetheless, as the cause and effect relationships between genes and disease come to light, there will be a growing need for more provider education so that new medical school graduates and the average PCP in clinical practice are ready for this future.

Scott McGrath, with the School of Interdisciplinary Informatics at the University of Nebraska at Omaha, and his colleague outline a detailed plan to educate the profession. They include five primary recommendations:
- Make genetics a core competency.
- Introduce biomedical and bioinformatics tools to medical students.
- Adjust entry requirements for medical schools.
- Offer dedicated training tracks in medical schools.
- Integrate emerging research into graduate medical education.

To foster a deeper knowledge of genomics among practicing physicians, McGrath and Ghersi recommend offering more CME courses in genomics and setting up a certification system in precision medicine. Increasing the financial incentives would also encourage physicians in training to consider genetics as a specialty. Currently, a medical geneticist earns about $159,000 annually while ED doctors earn about $320,000 and pediatricians about $207,000.

Consider another reason why physicians spend little time discussing genetics with their patients or ordering genetic tests: It is estimated that there are 26,000 genetic tests available for over 5400 conditions, and yet a study that looked at how PCPs, cardiologists, and psychiatrists use pharmacogenomics testing found that only 15%–30% ordered these types of panels [12]. One reason for this lack of enthusiasm is the lack of support from the agencies and associations that publish national practice guidelines. Katherine Johnansen Taber, PhD, the director of the American Medical Association's Precision Medicine Program, explains the situation this way:

> Very few recommendations that are applied in the primary care setting include genetic testing. As an example, of the approximately 100 recommendations of the U.S. Preventive Services Task Force, only one, on BRCA-related cancer, includes a recommendation for genetic counseling and testing. In Healthy People 2020, only two genomics objectives are included: a recommendation for testing newly diagnosed colorectal cancer cases for Lynch syndrome, and a recommendation for genetic counseling for those with a family history of breast and ovarian cancer. Even the CDC's Evaluation of Genomic Applications in Practice and Prevention (EGAPP) initiative – specifically charged with evaluating the clinical situations in which genetic testing should occur – has recommended it in only two clinical scenarios…. It's not hard to imagine that with so few guidelines recommending testing, many physicians don't perceive a substantial clinical value from them.

Clearly then, a lack of knowledge of genetics and genomics is only part of the equation. And whether the paucity of recommendations to perform genetic testing is due to an overly conservative mindset among policy making groups or an unwillingness to recommend procedures that have too little evidence to justify their use is a debate that takes us beyond the scope of this book.

ADDRESSING THE WORKLOAD ISSUE

Embracing a personalized approach to patient care may require more time on the part of clinicians. If the Precision Medicine Initiative generates data on patients' genomes, numerous environmental risk factors such as stress responsiveness, nutritional status, exposure to pollutants, as well as data from mobile medical apps, and so on, this will require the attention of practitioners who are already coping with a long list of new regulations and demands on their time. Critics are therefore asking: How can we incorporate precision medicine into clinical practice when it will add to physicians' workload, overwhelming them with more checkboxes to fill in, and more variables to take into account?

We may be asking the wrong question. A more pertinent question would be: How can the workflow of the average clinician be *redesigned* to accommodate the introduction of precision medicine. Ralph Snyderman, MD, Carline Meade, BS, and Connor Drake, MPA, with the Duke University Center for Research on Personalized Health Care, Duke University, sum up the challenge: "The lofty goals for personalized and precision medicine are out of reach until the health-care delivery system is designed for health promotion, comprehensive disease prevention, and efficient adoption of PPM capabilities" [13].

Redesign is always a difficult process, and in health care it is even more difficult because the profession is too often stuck in the here and now, clinging to the status quo. A similar dilemma has developed since the adoption of electronic health record systems. Many practitioners have complained that electronic health records disrupt their workflow. Part of the solution to that problem is to redesign the workflow process. We need a similar paradigm shift in clinical care.

That shift is going to require providers to move away for the traditional approach to patient care, in which they typically react to an episode—initial acute symptoms in a previously healthy patient or a flare-up of an existing disorder—and begin to take a more proactive role, embracing many of the predictive tools we have discussed earlier in the book, and developing preventive and mitigating strategies. This approach must include risk assessment tools, biomarkers, and targeted therapies, some of which are already available. Of course, patients will have to buy into this model too.

At the core of this redesigned model is what Snyderman and his colleagues refer to as personalized health planning (PHP), which consists of five steps:

- Estimate each patient's interest in engagement and self-management, and their ability to participate in their own care (That will depend on their educational level, their motivation, their trust in the health-care system, and their family support). Synderman et al. recommend having patients complete a self-assessment that addresses their needs, preferences, and goals.

- Assess each patient's health status and risks by employing conventional, genomic, and other precision diagnostic tools.
- Arrive at a set of shared goals after consulting with the patient.
- Incorporate those goals into a personalized plan, making sure to include the patient in the planning process.
- Coordinate the patient's care, enlisting the help of others on the care team.

This PHP process has attracted the interest of the US Veterans Health Administration, which has launched pilot projects to test out the concept in a primary care setting. Pilot projects have been set up in five clinics in Boston and Bonham, Texas. In this setting, patients are managed using a holistic approach that even takes into account aspects of one's health that are rarely assessed in traditional medical settings, including spirituality, personal relationships, emotional health, and personal development [14].

There are also initiatives in the private sector that aim to redesign clinician workflow and shift it toward a more holistic, personalized model. Interpreta, for example, a health-care analytics firm founded by Ahmed Ghouri, MD, Gary Rayner, and Raghu Sugavanam, is designed to continuously update, interpret, and synchronize clinical and genomics data, with the goal of creating a personalized roadmap. Its real-time digital tools and dashboard provide physicians, care managers, and payers with the patient-specific information that can facilitate precision medicine.

Like several other forward-minded thinkers interested in advancing personalized medicine, Ghouri, Rayner, and Sugavanam concentrate their efforts on using the data they collect to generate risk scores that clinicians can use to redirect their day-to-day decisions. But Interpreta takes this concept to a higher level by putting real-time data into the hands of clinicians to help them recognize high disease burden and acute clinical needs. Its analytics engine identifies care gaps, poor medication adherence, and risk of hospitalization.

Another area in which personalized medicine is taking hold is diabetes care, which we have discussed earlier in the book. Although personalized medicine remains in its embryonic stages for many disorders, the promise of this new model is already being realized for many patients with diabetes as they use mobile apps to record blood glucose patterns and share them with their providers. With that in mind, it is probably no coincidence that the innovative thinkers at Duke University Medical Center have incorporated

PHP into their management of type 2 diabetes. PHP has been included within shared medical appointments for this patient population.[2]

Unfortunately the reimbursement incentives currently in place in the American health-care system are not conducive to this type of clinical care model. Unless we continue to move from fee-for service medicine to pay-for-value medicine, it is unlikely we will see a shift.

While we wait for the change in reimbursement incentives, it is still possible for clinicians to incorporate many elements of personalized medicine into everyday clinical practice. But that requires more Cuisinarts and more Roombas. We need labor-saving devices, namely more software and hardware tools that will reduce workload and allow machines to do more of the work.

A randomized controlled study conducted by the well-respected Joslin Diabetes Center in Boston is worth consideration in this context [15]. As most clinicians who care for type 2 diabetic patients know, one of the most difficult periods for patients is when they are required to add insulin to a treatment regimen that includes an oral agent like metformin. Titrating the dose is challenging for patients and physicians alike, and incorrect insulin use remains one of the main reasons patients end up in the emergency room. The experts at Joslin have devised a creative way to utilize mobile technology to assist with this transition, easing the burden on clinicians and empowering individual patients to take on more of the responsibility for adjusting their basal insulin dose.

As we discussed in Chapter 5, Joslin's clinicians employed a cloud-based management system that uses an individualized treatment plan that patients access on their tablet, with a wireless glucose meter synched to the system to transmit glucose readings. The point we want to emphasize here is that practitioners were able to spend *less* time with the experimental group, about 66 versus 82 min/patient.

Clinical decision support systems are another labor-saving "device" that can lighten the workload of practitioners interested in personalized medicine—if they are configured correctly. For example, pharmacogenomic testing has been shown to help individualize drug therapy, reducing the risk of adverse effects by identifying slow and fast metabolizers. Since most PCPs do not have the time or expertise to locate all the relevant gene/drug interactions that could affect their patients' response

---

[2] Shared medical appointments typically involve gathering several patients together for an appointment in the same clinical setting and with several clinicians participating in each appointment. Participants may include a nurse or psychologist with patient education skills, a prescribing practitioner, and others.

to medication, they need quick access to a database with the relevant facts. Several initiatives have been undertaken to fold this type of data in clinical decision support programs. But these initiatives themselves pose several challenges.

The database has to limit itself to the most clinically relevant gene/drug interactions. It needs to streamline the presentation of information so that clinicians only see what they need to make a diagnostic or therapeutic decision, and it has to be configured in a way that would avoid alert fatigue. But even if the database met all those requirements, it still would not gain much traction in clinical practice if there was no reimbursement mechanism that pays for specific pharmacogenomic testing. And the schizophrenic positions taken by various government agencies make payment for these tests problematic.

For example, FDA says such testing can help identify responders and nonresponders to medications, avoiding adverse events, and optimizing drug dosing. But the Centers for Medicare and Medicaid Services has eliminated coverage of most genetically based drug sensitivity tests. Federal regulations require CMS to only pay for laboratory procedures that it deems "reasonable and necessary" and that does not include most of gene/drug interactions that have been confirmed by scientists—with the exception of CYP2C19 testing for clopidogrel. Commercial insurers typically follow CMS guidelines. For instance, BlueCross Blue Shield of North Carolina has a corporate medical policy for "Pharmacogenetic Testing for Drug Metabolism" that states "Pharmacogenetic testing for drug metabolism is considered investigational. BCBSNC does not provide coverage for investigational services" [16]. Many experts would disagree, as do some research studies.

There are more than 160 drugs approved from the FDA that include recommendations on adjusting dosing based on an individual patient's genetic profile. How does one adjust the dose of these medications if there is no third party payer to cover the cost of testing? These tests can produce benefits. For example, Fagerness et al. have compared psychiatric patients who had pharmacogenetic testing done (n = 111) to those who did not (n = 232) and found the testing resulted in better adherence to the drug regimen and saved $562 over a 4 month follow-up period [17]. Pharmacogenetic testing has also proven valuable to test cancer patients for the HER2 gene to determine if they will respond to trastuzumab (Herceptin). There is also evidence to support the need to test for genetic variants of the cytochrome P450 (CYP)2C9 system and the gene for

VKORC1, both of which have been are associated with differences in warfarin dose requirements. One analysis found that testing for both genes "would decrease health care spending in the United States by $1.1 billion annually with a range of $100 million to $2 billion" [18]. On the other hand, research to determine if pharmacogenetic testing would help predict patients' response to antidepressants or their tolerance to treatment of major depression has been less encouraging, suggesting that it is neither cost-effective nor clinically efficacious [19].

IS THERE ADEQUATE SCIENTIFIC EVIDENCE?

Throughout this book, we have cited numerous clinical studies and data analyses and outlined the conceptual underpinning in support of precision/personalized medicine. The evidence is stronger in some specialties than others. Diabetes remains one of the more well-supported areas. But even in this clinical area, there are strengths and weaknesses in the evidence. While mobile apps are now making individualized diabetes care a reality, there is limited genomic data available to assist in this personalization.

The search for genetic variants and biomarkers that would allow clinicians to predict which individuals will develop type 2 diabetes, or which patients would progress more rapidly, have been disappointing. Genome-wide association studies have tried to detect significant links between the disease and specific mutations with little success. Similarly, genetic risk scores have been developed that combine data from several genetic variants associated with the disease but these too have not been able to predict type 2 diabetes in a way that would have practical value in clinical practice. For example, a genetic risk score that included 18 genetic variants has been developed using data from the Framingham Heart Study. Its predictive ability was no better than predictions gleaned from traditional nongenetic risk factors [20]. Considering the fact that the cost of collecting traditional risk factors is less expensive than performing specialized gene testing, it is hard to justify the expense.

Since most cases of diabetes are polygenic and environmental in nature, genetic markers have yet to have a real impact. But on a more positive note, genetic analysis is proving useful in detecting single gene forms of the disease. For example, HNF1A mutations have been discovered in maturity onset diabetes of the young (MODY); KCNJ11 mutations have been linked to neonatal diabetes. Awareness of these mutations has practical application in patient care.

ADDITIONAL BARRIERS TO SURMOUNT

Another significant roadblock to personalized medicine is lack of enthusiasm among many patients. Without adequate patient engagement, there is only so much any clinician can do to individualize treatment. Among the issues that prevent full patient engagement:

- Lack of a sense of personal responsibility
- Lack of aptitude to grasp the concepts or understand the need to collect data
- Suspicion about the value of medicine and preventive measures. Many Americans still do not believe preventive measures will have a real impact—"After all my grandfather smoked 2 packs of cigarettes a day and never got cancer." Some even go as far as to imagine that preventive advice is an attempt by the government to control their lives.

The last chapter of this book will deal with this issue in more detail.

Finally patient privacy and data security remain a concern as the nation moves toward a personalized medicine model. Chapter 9 deals with those concerns.

REFERENCES

[1] U.S. Food and Drug Administration. Table of pharmacogenomic biomarkers in drug labeling. http://www.fda.gov/Drugs/ScienceResearch/ResearchAreas/Pharmacogenetics/ucm083378.htm.

[2] Tannock IF, Hickman JA. Limits to personalized cancer medicine. N Engl J Med 2016;375:1289–94.

[3] Bower H, Bjorkholm M, Dickman PW, et al. Life expectancy of patients with chronic myeloid leukemia approaches the life expectancy of the general population. J Clin Oncol 2016;34:2851–7.

[4] Ohm L, Lundqvist A, Dickman P, et al. Real-world cost-effectiveness in chronic myeloid leukemia: the price of success during four decades of development from non-targeted treatment to imatinib. Leuk Lymphoma 2015;56:1385–91.

[5] Swain SM, Baselga J, Kim S-B, et al. Pertuzumab, trastuzumab, and docetaxel in HER2-positive metastatic breast cancer. N Engl J Med 2015;372:724–34.

[6] Fojo T, Mailankody S, Lo A. Unintended consequences of expensive cancer therapeutics—the pursuit of marginal indications and a me-too mentality that stifles innovation and creativity: the John Conley lecture. JAMA Otolaryngol Head Neck Surg 2014;140(12):1225–36. http://dx.doi.org/10.1001/jamaoto.2014.1570.

[7] Howard DH, Bach PB, Brendt ER, et al. Pricing in the market for anticancer drugs. J Econ Perspect 2015;29(1):139–62.

[8] Ferkol T, Quinton P. Precision medicine: at what price? Am J Respir Crit Care Med 2015;192:658–9. http://www.atsjournals.org/doi/full/10.1164/rccm.201507-1428ED#readcube-epdf.

[9] Haslem DS, Van Norman C, Fulde G, et al. A retrospective analysis of precision medicine outcomes in patients with advanced cancer reveals improved progression free survival without increased health care costs. J Oncol Pract September 6, 2016. pii:JOPR011486. [Epub ahead of print].

[10] Elliott LS, Henderson JC, Neredilek MB, et al. Clinical impact of pharmacogenetic profiling with a clinical decision support tool in polypharmacy home health patients: a prospective pilot randomized controlled trial. PLoS One 2017;12(2):e170905.

[11] McGrath S, Ghersi D. Building towards precision medicine: empowering medical professionals for the next revolution. BMC Med Genom May 10, 2016;9:23. http://dx.doi.org/10.1186/s12920-016-0183-8. https://bmcmedgenomics.biomedcentral.com/articles/10.1186/s12920-016-0183-8.

[12] Taber KJ. Why do physicians experience genomics education gaps, and what can we do about it? J Precis Med 2015;1(May/June):76–80. http://www.thejournalofprecision-medicine.com/archive-manager/test11/.

[13] Snyderman R, Meade C, Drake C. To adopt precision medicine, redesign clinical care. NEJM Catal February 5, 2017. http://catalyst.nejm.org/adopt-precision-medicine-personalized-health/.

[14] Simmons LA, Drake CD, Gaudet TW, et al. Personalized health planning in primary care settings. Fed Pract 2016;33(1):27–34.

[15] Hsu WC, Lau KH, Huang R, et al. Utilization of a cloud-based diabetes management Program for insulin initiation and titration enables collaborative decision making between healthcare providers and patients. Diabetes Technol Ther 2016;18:59–67.

[16] BlueCross BlueShield of North Carolina. Corporate medical policy: pharmacogenetic testing for drug metabolism. July 2016. https://www.bcbsnc.com/assets/services/public/pdfs/medicalpolicy/pharmacogenetic_testing_for_drug_metabolism.pdf.

[17] Fagerness J, Fonseca E, Hess GP, et al. Pharmacogenetic-guided psychiatric intervention associated with increased adherence and cost savings. Am J Manag Care 2014;20:e146–156.

[18] Wu AC, Fuhlbrigge. Economic evaluation of pharmacogenetic tests. Clin Pharmacol Ther 2008;84:272–4.

[19] Rosenblat JD, Lee Y, McIntyre RS. Does pharmacogenomic testing improve clinical outcomes for major depressive disorder? A systematic review of clinical trials and cost-effectiveness studies. J Clin Psychiatry January 3, 2017. epub https://www.ncbi.nlm.nih.gov/labs/articles/28068459/.

[20] Floyd JS, Psaty BM. The application of genomics in diabetes: barriers to discovery and implementation. Diabetes Care 2016;39:1858–69.

CHAPTER EIGHT

Interoperability and Personalized Patient Care

Interoperability is the glue that holds precision medicine together. It is essential for the success of the US federally funded Precision Medicine Initiative (PMI) and it is important for the success of numerous nongovernment sponsored precision medicine projects.

The proposed budget for PMI included $5 million for the Office of the National Coordinator of Health Information Technology (ONC), a division of the US Department of Health and Human Services. The funds are specifically intended "To support the development of interoperability standards and requirements that address privacy and enable secure exchange of data across systems" [1]. Without the interoperability of EHRs, for instance, it would be virtually impossible to gain access to the clinical parameters used to detect risk factors and individual susceptibilities to a variety of diseases. It would be impossible to collect the genotypic and phenotypic data and the diagnostic data that are so essential to arriving at an individualized treatment plan.

DEFINING THE BASICS AND FOLLOWING A ROADMAP

ONC published *Connecting Health and Care for the Nation: A Shared Nationwide Interoperability Roadmap—Version 1.0 Final (Road map)* in 2015 to help turn the potential of precision medicine into a reality. The document explains that, despite the progress made with the enactment of the Health Information Technology for Economic and Clinical Health Act, "2015's interoperability experience remains a work in progress" [2]. Many clinicians have been far less charitable in their assessment and prefer to think of the current situation as a "disaster in situ." But regardless of one's assessment of the current state of interoperability, the ONC roadmap provides a clear direction, which if followed, will likely create "a learning health system where individuals are at the center of their care; where providers have a seamless ability to securely access and use health information from different

Realizing the Promise of Precision Medicine
ISBN 978-0-12-811635-7
http://dx.doi.org/10.1016/B978-0-12-811635-7.00008-7

sources; where an individual's health information is not limited to what is stored in electronic health records (EHRs) but includes information from many different sources (including technologies that individuals use) and portrays a longitudinal picture of their health, not just episodes of care; where diagnostic tests are only repeated when necessary, because the information is readily available; and where public health agencies and researchers can rapidly learn, develop, and deliver cutting edge treatments."

The Healthcare Information and Management Systems Society (HIMSS) explains that interoperability is "the extent to which systems and devices can exchange data, and interpret that shared data. For two systems to be interoperable, they must be able to exchange data and subsequently present that data such that it can be understood by a user" [3]. To make this exchange of data seamless is going to require the cooperation of several stakeholders and a well-structured plan of attack. The ONC roadmap lays out this plan of attack by establishing three time-sensitive goals and four critical pathways, and it spells out the plan by dividing the challenges into three sections: Drivers, policies and technical component, and outcomes.

- From 2015 to 2017, the goal is to have an information system capable of sending, receiving, finding, and using priority data domains to improve health-care quality and outcomes.
- 2018–20: Expand data sources and users in the interoperable health IT ecosystem to improve health and lower costs.
- 2021–24: Achieve nationwide interoperability to enable a learning health system, with the person at the center of a system that can continuously improve care, public health, and science through real-time data access.

The four immediate pathways that need attention are outlined in the Roadmap document:

- Improve technical standards and implementation guidance for priority data domains and associated elements. In the near-term, the Roadmap focuses on using commonly available standards, while pushing for greater implementation consistency and innovation associated with new standards and technology approaches, such as the use of application programming interfaces (APIs).
- Rapidly shift and align federal, state, and commercial payment policies from fee-for-service to value-based models to stimulate the demand for interoperability.
- Clarify and align federal and state privacy and security requirements that enable interoperability.

- Coordinate among stakeholders to promote and align consistent policies and business practices that support interoperability and address those that impede interoperability.

As item #1 points out, ONC realizes that value of application programming interfaces to facilitate interoperability. An API is a set of digital tools or instructions to help two pieces of software communicate. If, for example, one wants to develop an iPhone app, one must communicate in a specific way with iOS, the operating system on Apple devices. For that, one will need access to an Apple API. Similarly, proprietary EHR systems can be made to communicate much more readily when open-source APIs are made available to link those EHR systems to one another and to other sources of clinical, administrative, and financial data. APIs can also ease the way for third party developers to create applications that allow mobile monitoring devices to communicate with the digital records systems that physicians currently use. That in turn makes it easier for a patient using a heart rate monitoring app to transmit their readings to their provider, for example. As we have discussed in earlier chapters, such patient-generated data will play a key role in personalized medicine.

WHAT DRIVES INTEROPERABILITY?

Even a superficial look at the statistics on health information exchange (HIE) demonstrates the urgent need for APIs and other creative solutions. At the time the ONC roadmap was published in 2015, about 41% of hospitals had electronic access to clinical information from outside providers or other sources when treating patients. In 2014, about 78% of hospitals were sending electronic summary of care documents out and 56% were receiving them. Despite this progress, fewer than 50% of hospitals were integrating these data into patient records, and in 2013, only 14% of physician offices were sharing electronic patient data with other providers outside their own facility. The term "work in progress" certainly seems charitable in this context.

To meet these challenges, the Interoperability Roadmap divided its mission into three sections: *Drivers, policy and technical components*, and *outcomes*, as Fig. 8.1 illustrates. The drivers are the incentives needed to push interoperability forward. *Policies and technical components* are self-explanatory, and *outcomes* refers to the measurement framework that needs to be put in place to demonstrate that we move from a work in progress to a fully mature learning health system.

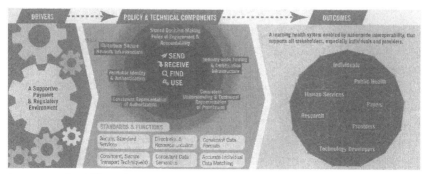

Figure 8.1 The Office of the Coordinator of Healthcare Information Technology has organized the quest for interoperability into three categories: drivers, policy and technological components, and outcomes.

The federal government has chosen not to mandate nationwide interoperability. Instead it has created several persuasive incentives that will encourage adaption of the policies, standards, and implementation specifications recommended in the Roadmap. The Medicare Access and CHIP Reauthorization Act (MACRA) is one of the most convincing incentives.

MACRA, which was enacted in 2015, was designed to move the healthcare industry from a fee-for-service to a value-based model. It requires practitioners who care for Medicare patients to demonstrate that they are offering quality care by reporting a variety of quality metrics. It is certainly no coincidence that the quality reporting process is heavily dependent on information technology—and on better interoperability. The two components of MACRA, the Merit-based Incentive Payment System (MIPS) and the more ambitious alternative payment models (APMs), require clinicians to use certified technology to coordinate care. And since MIPS and APMs usually require providers to work hand in hand with one another in various practice settings and in diverse locations, it would be virtually impossible to collect the needed metrics and demonstrate quality care without seamless data exchange.

MIPS takes the place of the Meaningful Use of EHR Incentive Program, Physicians Quality Reporting System, and Value Modifier Program. That does not mean the electronic reporting requirements of these older programs go away. They are more or less "tucked" into MIPS, which means it will still be necessary to meet a long list of reporting requirements that are almost impossible to navigate without digital tools that talk to one another.

Consider the 2017 MIPS measures and activities. Most participants are expected to report up to six quality measures, including an outcome measure, for a minimum of 90 days [4]. Providers can choose from a long list of quality measures to find those that best suit their patient population and their strengths. Let us assume you choose breast cancer screening as one of the metrics. You are then expected to report the percentage of women 50–74 years of age in your practice who had a mammogram to screen for the disease. Among the ways the practice can submit these data to the Centers for Medicare and Medicaid Services: through an EHR, through claims or a registry, or directly on the CMS web interface. A second metric you might choose is "Melanoma: Coordination of care." To meet that quality measure, one must report the percentage of patient visits, regardless of age, with a new occurrence of melanoma who have a treatment plan documented in the chart that was communicated to the physician providing continuing care within 1 month of diagnosis. Metrics like these seem straightforward but there are myriad ways in which the data collection can go awry.

In a group practice spread out over numerous locations, if one or more offices are not using the same certified EHR system, collecting and collating these statistics will be difficult if the systems are not using the same computer languages to talk to one another or the same transport protocols. Similarly, if the quality metrics you choose require interacting with lab partners, referring physicians, public health agencies, nursing homes, clinicians using mobile messaging systems, or patients using remote sensors, you face a tangled web of conflicting software suites that were not initially designed to communicate with one another. Without a common set of interoperability standards, which are outlined in the 2017 Interoperability Standards Advisory (ISA), you are faced with a technological Tower of Babel.

A sampling of several quality metrics will make it even more obvious that practitioners will not survive very long without better interoperability. Consider some of the quality metrics listed in the "Advancing Care Information" component of MIPS [5].

- Secure Messaging: For at least one unique patient seen by the MIPS eligible clinician during the performance period, a secure message was sent using the electronic messaging function of certified EHR technology to the patient (or the patient-authorized representative), or in response to a secure message sent by the patient (or the patient-authorized representative), or in response to a secure message sent by the patient (or the patient-authorized representative).
- Immunization Registry Reporting: The MIPS eligible clinician is in active engagement with a public health agency to submit immunization data and receive immunization forecasts and histories from the public health immunization registry/immunization information system.

- Send a Summary of Care: For at least one transition of care or referral, the MIPS eligible clinician that transitions or refers their patient to another setting of care or health-care provider: (1) creates a summary of care record using certified EHR technology; and (2) electronically exchanges the summary of care record.

POLICY AND TECHNICAL COMPONENTS

Incentives are important, but once they begin driving providers to improve the process of HIE, the next step is establishing a common set of standards, services, policies, and practices—and establishing a governance process to put these things in place.

In December 2016, ONC published the 2017 ISA that spells out the vocabulary, code sets, terminology standards, and implementation specifications it recommends but does not mandate. The fact that the advisory is not a static document but a collection of standards that can change over time as the market dictates is one of its advantages. Had the Advisory been a set of regulations cemented in time, interoperability would likely suffer.

One of its primary purposes is to "provide the industry with a single, public list of the standards and implementation specifications that can best be used to address specific clinical health information interoperability needs. Currently, the ISA is focused on interoperability for sharing information between entities and not on intraorganizational uses." A few examples from the Advisory illustrate its value.

If one wants to communicate information about a patient's allergy reactions and smoking status, the Advisory recommends using LOINC and SNOWMED-CT. To communicate details on immunizations, there is HL7 Standard Code Set CVX—Clinical Vaccines Administered and HL7 Standard Code Set MVX-Manufacturing Vaccine Formulation; to represent analytic data for research purposes, Clinical Data Interchange Standards (CDISC) Controlled Terminology for Regulatory Standards hosted by NCI-EVS and CDISC Controlled Terminology for CDISC Therapeutic Area Standards Hosted by NCI-EVS.

Elsewhere the Advisory provides recommendations of higher level data reporting including care plans, clinical decision support, clinical quality reporting, diet and nutrition, patient education materials, electronic prescribing, and summary care records. Among the standards and implementation specifications: HL7 Clinical Document Architecture (CDA), Release 2.0, Final Edition;

- HL7 Fast Healthcare Interoperability Resources (FHIR) Profile: Quality (QI Core), DSTU Release 1;
- HL7 Version 3 Standard: Diet and Nutrition, STU Release 1;
- Digital Imaging and Communications in Medicine; and
- National Council for Prescription Drug Programs SCRIPT Standard, Implementation Guide, Version 10.6.

The CDA provides a standardized way for health–care providers to create electronic clinical documents that are readable by humans and machines. It provides structure, coding, a semantic framework, and a markup language for the creation of these clinical documents.

CDA documents use XML as a markup language and use templates that offer standardized groupings of information. The architecture can be used to construct continuity of care documents (CCDs), diagnostic imaging reports, discharge summaries, operative notes, and other clinically relevant documents that need to be sent from one provider to another.

The Interoperability Roadmap also addresses governance in its discussion of electronic health information sharing arrangements. As it points out, there are several preexisting arrangements for sharing data, including HIEs, networks, and trust communities. Several other arrangements between unaffiliated and/or competing groups have also arisen recently, including CommonWell Health Alliance, the Sequoia Project, and DirectTrust. The Roadmap's guarded optimism about such initiatives is summarized here:

Despite the potential and intention of existing electronic health information sharing arrangements, they differ from each other in fundamental ways that make it difficult for them to work together. They often have differing immediate goals and differing methods or standards to achieve those goals. Some networks that support health care, implement information sharing arrangements through formal contracts or legal data sharing and use agreements, while some rely on self-attestation or independent accreditation. Some operate technical testing programs while others do not. And most, but not all, operate some level of technical infrastructure. The result can be a complex web of electronic health information sharing arrangements that creates some degree of interoperability within specific geographic regions, organizational and technology developer boundaries, but fail to produce seamless nationwide interoperability to support a learning health system. These existing arrangements, that are often one-to-one contracts or data use agreements, are unlikely to scale nationwide and may not be extensible to new, novel data uses that support health….

Moving forward successfully with shared decision-making, rules of engagement and accountability requires an understanding of what has been tried in the past.

ONC has made several attempts to overcome variation across existing electronic health information sharing arrangements to enable nationwide interoperability, but with limited success…. ONC continues to believe that the electronic health information sharing arrangements described above are valuable tools to promote interoperability among unaffiliated organizations. However, there are evident gaps, overlaps and conflicting approaches among and between the various organizations that prevent the sharing of electronic health information. Reaching the near- and long-term nationwide interoperability goals will require existing arrangements to be able to share information across their respective boundaries, even between competitors, and should focus on the Roadmap's near term goal of sending, receiving, finding and using priority data domains.

LEARNING FROM OUR MISTAKES

Since the ONC Roadmap was published, much progress has been made to address the aforementioned obstacles to mature, nationwide interoperability. Looking back at how electronic health-care interoperability was born can provide some insights into how it has progressed in the last few years.

In 2004, George W. Bush decided to appoint a national health-care coordinator for information technology and chose David Brailer as his first ONC leader. Bush's goal was that within 10 years: "Every American must have a personal electronic record… The federal government has got to take the lead in order to make this happen" [6]. Unfortunately, ONC was given no funds to accomplish this goal. The original $50 million budget slated for ONC was canceled by Congress, forcing HHS to find funds from within its department.

Brailer's bully pulpit allowed him to emphasize the urgent need for IT standards. He created the Health Information Technology Standards Panel and appointed one of us (John Halamka) as chairperson. The panel was convened as a public/private partnership by the American National Standards Institute, which promulgates open standards. One criticism was that the Panel attracted several large corporations that had a vested interest in maintaining the status quo in information technology, rather than attracting innovators eager to move health care in a direction that would encourage universal interoperability. The result was that during the Bush administration, the Panel created a variety of standards that protected the status quo. They included many standards from the 1990s—hardly innovative technology. Unfortunately, the only stakeholders who could afford to implement these standards where the large corporations that had encouraged their creation.

When the Obama administration took the reins, the agenda shifted. The new administration used the Federal Advisory Committee Act (FACA) to reshape the standards development process. The idea was to assemble experts from the community to provide advice on the subject. FACA mandated a new level of openness and transparency, including a much more diverse set of stakeholders. They included patient advocacy groups, children's groups, gay community representatives, family advocacy groups, and others. Unfortunately, none of these groups had any expertise in creating the complex standards required to move EHR interoperability forward. Despite these roadblocks, this period did result in significant advances in the development of health-care IT standards, including laying the groundwork for Fast Healthcare Interoperability Resources (FHIR).

During this period, progress was also made in defining a mature IT standard. In the past, an ideal, mature standard was one that the large IT-dependent corporations wanted to maintain. The redefined mature standard, which has been outlined in a 2015 paper in the *Journal of the American Medical Informatics Association*, was less expensive, easily accessible, well documented, and easy to use [7].

Despite pushback from EHR vendors, these standards gradually gained traction in the health IT community. However, during this period, the standards suffered from "platypus syndrome." Because there was so much emphasis on arriving at consensus, the standards tried to be all things to all people, so they looked like a platypus, which seems to have been constructed from parts of several unrelated species. The resulting IT constructs suffered from too much optionality. They tried to serve too many stakeholders and too many use cases. Put another way, optionality means giving users too much choice. By way of analogy, if one is trying to determine a standard for driving automobiles on public roads, it means giving drivers the option of driving on the left *or* right side of the road. But in practice, it means drivers will drive on the left *and* right side of the road—not the safest way to run a highway system.

To remedy the situation, the Obama administration hired JASON, a group of elite scientists, to advise it on a variety of secret and public projects, including health-care data. In 2013, the Agency for Healthcare Research and Quality published a report prepared by JASON that trashed most of the current interoperability standards. The group had interviewed Google, Amazon, Facebook, and others and found that they all had standards that can be easily implemented with very little training [8].

That report generated controversy and debate within the federal government and eventually led to recommendations from the President's Council of Advisors on Science and Technology to simplify the standards and make them more user friendly. And in response to the JASON report, like-minded

IT experts decided to form the Argonaut Project, the goal of which was to simplify the interoperability standards, to use the same approach that every other industry uses, and to take a page from the likes of Google, Amazon, and other successful companies. The strategy that the Argonaut Project recommended was to avoid complicated health-care specific standards; instead it recommended the use of JAVA script object notation (JSON), which is the way every web-based company shares data and the way software engineers build add-ons, modules, and so on.

JSON is a very simple syntax. If a user submits a few data elements to a website, they will get back the same data elements and they are always the same elements. If the JSON application programming interface is written so that an allergy has five data elements, it will always have five data elements. FHIR follows this same approach, using JSON and simple web-based calls to enable developers to very rapidly consume these data. And the fact that there is no optionality makes the process seamless.

A good example of the simplified data exchange method is a mobile app created by Beth Israel Deaconess Medical Center in Boston called BIDMC@home. It allows patients at the hospital or affiliated physician office to transfer their medical data from one location to another, including interstate transfers. This bypasses the problems that occur when health-care organizations attempt to send patient data across state lines. Since each state has its own set of laws, which preempt the federal HIPAA rules, such transfers can get exceedingly complicated.

BIDMC@home, which has a FHIR interface, can be installed on a patient's phone, and can receive all their records. Then they can send that data to a hospital in Philadelphia, which will have a similar app capable of receiving and understanding the content. The other advantage of giving the data to the patient directly is that the Office of Civil Rights says it will never hold a provider organization responsible for shared medical data if it comes from the patient.

Of course, like all interoperability standards, FHIR has its weaknesses. FHIR is very early in its development. In its current draft version, FHIR only has about 150 data elements. A typical electronic patient record may contain more than 1000 data elements so it will take time before it is fully embraced by the EHR vendor community.

ADDRESSING SECURITY CONCERNS

While making data exchange easier is essential to the precision medicine movement, so is security. The ONC's Interoperability Roadmap addresses this issue as well. To meet the policy and technical components of mature interoperability will require a secure infrastructure, verifiable identity and authentication procedures, authorization protocols that delineate the role of each data user and their specific rights to access patient data, a testing and certification infrastructure, consistent data semantics and data formats, secure transport techniques, and more.

It may seem too obvious to say that interoperability requires identity proofing and authentication, but it is a legitimate concern for health-care organizations that communicate with a wide variety of vendors, large and small. Many providers have refused to exchange electronic health information or let others access their network without proper identification and authentication procedures. The risk of data breaches by unauthorized persons is simply too great. The four-step process recommended by the ONC roadmap includes:

- Identity Proofing
- Credentials—a password and user name, or a physical token embedded in a photo ID, for instance
- Authentication—the process of using one's credentials to gain access to the system. Some networks require two-factor authentication, in which one has to have a second form of credentials.
- Authorization—it is common for different members of a health-care organization to be authorized to view different types of data. There may be no reason for a medical assistant to view patients' financial records, for instance, or a lab technician to view a patient's medication history.

Chapter 9 will discuss the importance of security as it relates to precision medicine in more detail. While the need for a secure network is essential for any health-care organization that handles protected health information, that need is vastly increased when PHI is shared in the wider networks required to arrive at nationwide interoperability. As ONC points out "In an interoperable, interconnected health IT ecosystem, an intrusion in one system could allow intrusions in multiple other systems." This concern may threaten the trust on which nationwide interoperability rests. Large organizations with extensive financial resources have the ability to defend their patient records with sophisticated encryption systems, mobile device management platforms, and an army of security analysts but many smaller organizations have limited resources and a lack of appreciation for the security threats now facing the industry. That cultural blind spot must change for a nationwide data network to prosper.

Among the cybersecurity controls and best practices needed to safeguard PHI, as outlined in the ONC Roadmap:

- Create and maintain a security risk management program.
- Maintain contracts, such as Data Use Agreements, Memoranda of Understanding/Memoranda of Agreement, Interconnection Security Agreements, and Business Associate Agreements. These documents are typically contracts between two parties that are in addition to each party's own internal compliance documents such as HIPAA privacy and security policies and procedures. These documents will need to scale beyond bilateral contracts to support nationwide interoperability.
- Share threat information across organizations and develop mature incident response capabilities.
- Perform operational and behavioral monitoring of user credentials, administrator credentials, and use of system credentials, particularly those credentials that have system-level access to APIs or databases that contain ePHI or individually identifiable health information.
- Ensure that health IT is developed and deployed securely, following Department of Homeland Security and National Institute of Standards and Technology guidance for building security into health IT products, not just putting products behind a secure exterior.
- Assess the security of applications and infrastructure via penetration testing, potentially conducted by third party experts, to identify vulnerabilities before they are exploited. (Penetration testing and related security measures will be discussed in Chapter 9).
- Encrypt the contents of all network messages in transit even if it is not legally required.
- Secure all data stored in any database connected to the network, whether through a companion system, interface engine or gateway, by encrypting data at rest and securing the encryption keys.
- Participate in bug bounty programs.

UNDERSTANDING THE STANDARDS THAT FACILITATE INTEROPERABILITY

One of the problems with the English language, and all other human languages, is that it is often not precise enough for computers to use when they communicate information. There are so many ambiguities, metaphors, and colloquial expressions that it is almost impossible to interpret sentences without the intervention of another human brain that is familiar with all the complexities. To cope with these challenges, technologists have developed a variety of standards, including computer-readable vocabularies, formatting systems, transport protocols, and security standards. Fig. 8.2, derived from the ONC Roadmap, illustrates these standards and how they are used.

We have already discussed several of the vocabularies used in health-care IT, including SNOWMED-CT, RxNorm, and LOINC. They provide the "words" needed to communicate clinical, administrative, and financial data from one network to another. Ideally, we want sending hospital A to use the same set of vocabularies as receiving hospital B. When they do, they have "semantic interoperability." Similarly, we want them to use the same formatting standards, the equivalent of syntax in human language so that they have "syntactic interoperability." In other words, we want both the sending and receiving hospitals to use the same "packaging." Consolidated clinical document architecture (C-CDA) is one of the packaging formats that ONC recommends. It lets the second hospital easily "unpack" the data and integrate it into its own computer systems without time-consuming human intervention. In this context, computers talking to computers are far more cost-effective than computers talking to humans who must then convert the conversation into a different computer friendly language.

CDA is used to create a variety of documents, including imaging reports, CCDs, and procedural notes. The formatting provided by CDA might be compared to the formatting in a word-processing application such as Microsoft Word. The text you type into Word is only part of the document; it also contains formatting that specifies the font size and font style you prefer, where the bullet points go, the color of the type, and so on. If

CATEGORIES OF STANDARDS	FUNCTIONS OF STANDARDS	EXAMPLES OF REAL WORLD USE OF THE STANDARDS
VOCABULARY & CODE SETS (SEMANTICS)	The information is universally understood	RxNorm Code for Ibuprofen is 5640
FORMAT, CONTENT & STRUCTURE	Information is in the appropriate format	C-CDA packages up data in the appropriate format
TRANSPORT	The information moves from point A to point B	SMTP and S/MIME to send the C-CDA from one setting to another
SECURITY	The information is securely accessed and moved	X.509: to ensure it is securely transmitted to the intended recipient
SERVICES	Provides additional functionality so that information exchange can occur	DNS+LDAP: to find the recipient's X.509 certificate to encrypt a message

Figure 8.2 The standards needed to achieve universal interoperability fall into five broad categories. The functions of each type of standard and examples in each category are illustrated above. *C-CDA*, Consolidated clinical document architecture; *SMTP*, Simple Mail Transfer Protocol; *S/MIME*, Secure/Multipurpose Internet Mail Extensions; *DNS*, Domain Name System.

you decide to share the Word doc with someone else, the computer coding that represents the fonts and other features travels along with the text and must be readable by the application that opens it next. If it cannot read Word formatting, the next user will see several errors on their screen.

Assuming two hospitals are using the proper vocabularies and formatting, they then must use compatible transport and security standards. The recommended standards for transporting data from one location to another include Simple Mail Transfer Protocol and Secure/Multipurpose Internet Mail Extensions to send C-CDA. Security standards include X-509.

WHAT OUTCOMES CAN BE EXPECTED?

Finally, there are four outcome goals discussed in the Roadmap. If interoperability is performed as it should, we should expect to see

- Patients and consumers can have full access to the electronic health records, have the ability to contribute to those records, and be able to send it to any electronic location they choose.
- Providers sharing patient information in a timely manner with patients and with other health professionals.
- Providers changing workflow processes to improve care coordination, including the closure of referral loops.
- Tangible progress is achieving better care, smarter spending, and a healthier population.

REFERENCES

[1] Precision Medicine Initiative Working Group. The precision medicine initiative cohort program – building a research foundation for 21st century medicine. September 2015.
[2] Office of the National Coordinator for Health Information Technology. Connecting health and care for the nation: a shared nationwide interoperability roadmap FINAL version 1.0. https://www.healthit.gov/sites/default/files/hie-interoperability/nationwide-interoperability-roadmap-final-version-1.0.pdf.
[3] Healthcare Information and Management Systems Society. What is interoperability? http://www.himss.org/library/interoperability-standards/what-is-interoperability.
[4] Department of Health and Human Services Quality Payment Program. MIPS overview. https://qpp.cms.gov/measures/performance.
[5] Department of Health and Human Services Quality Payment Program. Advancing care information. https://qpp.cms.gov/measures/aci.
[6] Conn J. 10 years after the revolution: health IT coordinators look back at the nation's progress. Modern Healthcare; April 5, 2014. http://www.modernhealthcare.com/article/20140405/MAGAZINE/304059980.
[7] Baker DB, Perlin JB, Halamka J. Evaluating and classifying the readiness of technology specifications for national standardization. JAMIA 2015;22:738–43.
[8] Agency for Healthcare Research and Quality. A robust health data infrastructure. 2013. https://www.healthit.gov/sites/default/files/ptp13-700hhs_white.pdf.

CHAPTER NINE

Privacy and Security

The practice of personalized medicine requires respect for the person receiving the individualized care. One way that respect is made manifest is by keeping patient data private and secure. That mandate applies to hospitals and medical offices providing such care just as much as it applies to the multimillion dollar Precision Medicine Initiative (PMI) being sponsored by the US government.

With these two scenarios in mind, we will first outline the general principles and best practices to safeguard protected health information (PHI), and then discuss the application of these principles and best practices to the PMI.

KEEPING PATIENT DATA SAFE AT THE GRASSROOTS LEVEL

The lofty goals of the PMI require security protocols to maintain PHI secure, but the sensitive information being managed in community practices and hospitals across the nation is just as important. It will be virtually impossible to gain the public's trust to commit to a national project unless they are convinced that their local providers can keep their data out of the hands of cyber thieves and nosey staff members.

One of the reasons some providers do not invest the necessary financial and staff resources to bolster the security of patient data is they do not believe there is a business case for doing so. They assume that the data are safe enough and the cost of tightening security would be too high.

Statistics bear out the fact that many health-care executives believe that there are other fiscal priorities that need to come before investment in stronger cybersecurity. For example, a recent survey conducted by the Healthcare Information Management Systems Society found only 64% of hospitals and medical practices have put encryption software in place to protect patient data as it is transported from one location to another [1]. Similarly, a survey conducted by the Ponemon Institute, a research center focused on data security, found that 73% of health-care organizations have

Realizing the Promise of Precision Medicine
ISBN 978-0-12-811635-7
http://dx.doi.org/10.1016/B978-0-12-811635-7.00009-9
163

yet to implement the necessary resources to prevent data breaches or detect them once they occurred [2]. A separate survey found that only 42% of health-care providers were planning to put encryption in place and only 44% are planning to set up single sign on and authentication on their web-based applications and portals [3].

These statistics strongly suggest that decision makers in the health-care community still see the need for more security as unwarranted. Some may even suspect that the call for more security is just an alarmist rant by information security specialists or vendors hoping to sell more software and hardware. That argument might stand up to scrutiny, were it not for the long list of data breaches that have been reported in the last few years—many of which were preventable.

The US Department of Health and Human Services (HHS) Office of Civil Rights (OCR) publishes a comprehensive list of health-care data breaches in the United States. It contains over 1000 breaches that affected 500 or more individuals. This so-called "Wall of Shame," which can be viewed at https://ocrportal.hhs.gov/ocr/breach/breach_report.jsf, includes some massive attacks, such as the one that compromised 78,800,000 individuals at the large medical insurer anthem—reported to the US Department of HHS on 3/14/13—or the breach that exposed 11,000,000 members of Premera Blue Cross (3/17/2015).

Several smaller organizations and individual clinicians have also been embarrassed by having their breaches posted on the site. Clinicians in Ohio, Texas, and California, for example, are included on the list by personal name, along with how many patient records were exposed in each facility and the type of breach that occurred, for example, theft, hacking, unauthorized access or disclosures, and/or improper disposal of records.

OCR is required by Section 13402(e)(4) of the Health Information Technology for Economic and Clinical Health Act to post any breach of unsecured PHI affecting 500 or more individuals. Even more disturbing for small medical practices and community hospitals is the fact that federal officials are now going after providers who have experienced PHI leakages that affect fewer than 500 individuals. In 2013, HHS announced that the Hospice of North Idaho had to pay $50,000 for violations of the Health Insurance Portability and Accountability Act (HIPAA) because the facility allowed an unencrypted laptop with PHI for 441 patients to be stolen. In the words of Leon Rodriguez, the Director of the OCR at the time: "This action sends a strong message to the health-care industry that, regardless of size, covered entities must take action and will be held accountable for safeguarding their patients' health information.... Encryption is an easy method for making lost information unusable, unreadable and undecipherable" [4].

In July, 2016, OCR began to not only investigate health-care organizations that have reported data breaches but also to catch delinquent providers off guard by relaunching a program that audits providers who have not reported any incidents. A pilot project that started in 2011–12 revealed several shortfalls. Mark Fulford, a partner at LBMC, an accounting and consulting firm in Brentword, TN, explains "The 2012 OCR audits revealed the health-care industry at large had not yet begun to take compliance seriously. An astounding two-thirds of audited entities had not even performed a complete and accurate risk assessment, which is the first step in putting a security strategy in place" [5].

That initial series of about 100 audits found that many providers had neither taken basic steps to protect their networks nor were they able to identify their vulnerabilities—an important requirement spelled out in the federal regulations. Some organizations did not even know where their PHI *resided*. And they could not say definitively what data had been stored in those mysterious locations. Adding insult to injury, OCR found many employees were accessing data from unsecured mobile devices in public locations. Similarly, the audits indicated that many health-care organizations were not training staff on how to manage PHI.

INSECURITY IS EXPENSIVE

If you are responsible for the financial welfare of your organization, no doubt one question that comes to mind is: How much will it cost me if I do not adequately safeguard our PHI? Although protecting patient information involves legal and ethical issues, let's just focus on the financial issues for the moment.

It is estimated that health-care organizations spend about $6 billion a year as a result of data breaches. Since that does not tell you much about the cost of a breach to an individual provider, one has to look more closely at specific expenses. If your patients' PHI is compromised and a federal investigation determines that the organization shares some of the responsibility for that data loss, expect each violation to cost between $100 and $50,000. That is per patient record. So a stolen laptop containing unencrypted records of 1000 patients can cost the practice between $100,000 and $1.5 million in penalties alone (Although $50,000 × 1000 = $50 million, the government caps these penalties at $1.5 million).

HHS provides more detail on how it calculates the fines, breaking them down into four categories. If HHS determines that you unknowingly allowed the data breach and had exercised reasonable diligence, the fine is still between $100 and $50,000 per violation. However, if the breach occurred due to a "reasonable cause," the range then jumps to $1000 to $50,000 per violation. A third category, for a breach resulting from willful neglect that was corrected in a timely manner, will result in a fine of $10,000–$50,000. And lastly, if your organization has willfully neglected to take precautions and did not correct the problem in a reasonable amount of time, the fine is at least $50,000 per violation, with a cap of $1.5 million per calendar year [6].

In addition to these broad criteria, numerous factors go into the HHS determination of how much to fine a health-care provider, including how much harm results from the violation and the facility's history of prior compliance with the HIPAA regulation. And although OCR is most interested in breaches of more than 500 patient records, the government will go after smaller incidents when they believe it serves the cause of justice, as mentioned above. In 2009, for instance, Massachusetts General Hospital (MGH) agreed to pay $1,000,000 to settle an HIPAA violation that only affected 192 patients. The OCR had MGH sign a resolution agreement requiring it to "develop and implement a comprehensive set of policies and procedures to safeguard the privacy of its patients." The agreement resulted from an OCR investigation that started with a complaint filed by a patient whose PHI was exposed. Since the 192 patients affected by the breach were being treated by Mass General's Infectious Disease Associates outpatient practice, which included patients with HIV/AIDS, the exposure of patients' data not only threatened to expose them to the possibility of identity theft but they also revealed their HIV status, clearly a very personal piece of information that most patients would want to keep confidential. And although the incident involved paper documents, the same judgment would likely have been made had this been an electronic breach [7].

Despite all the high profile cases in which government authorities have imposed heavy fines on health-care organizations, a recent analysis indicates that only a small percentage of providers who report breaches and found themselves on the federal "Wall of Shame" actually are fined. A recent report on more than 1140 large breaches from ProPublica, a nonprofit investigative journalism group, revealed that only 22 resulted in fines [8]. That translates into less than a 2% likelihood of being fined.

ADDITIONAL HEALTH INSURANCE PORTABILITY AND ACCOUNTABILITY ACT VIOLATION EXPENSES

A manager who is comfortable with taking risks might reason that a 2% risk is acceptable and provides no incentive to strengthen one's security protocols. That logic is faulty for several reasons.

Federal fines are only part of the expense an organization would incur should a PHI breach occur. You may also be responsible for having a forensic evaluation performed to determine how the breach happened. Assuming for the moment that your practice or hospital does not have the expertise and personnel to do this expert analysis, you may have to spend on average between $200 and $2000 per hour for third-party assistance [9].

Depending on the circumstances surrounding a data breach, you may also have to notify those patients and employees whose personal information has been exposed. That will likely cost up to $5 per notice.

Patients who have had their PHI exposed are entitled to some type of protection to reduce the risk of identity theft. According to a 2012 analysis from Zurich American Insurance Company, you can expect to pay $30 per patient per year to cover the cost of credit monitoring, identity monitoring, and restoration [9]. But that figure may be outdated and is likely to be higher now. An identity protection service like Lifelock costs about $110 per year retail, which would translate to $220,000 for 1000 patients over 2 years [10].

You also have to consider the cost of a legal defense. If the incident reaches the mass media, it is possible that you will face a class action lawsuit. On average that will cost an organization about $500,000 in lawyer fees and $1,000,000 for the settlement [9]. Of course, many cautious health-care executives would naturally think twice about informing the local media about a data breach, but the law does not give you a choice in the matter. The HIPPA breach notification rule states that following a breach of unsecured PHI involving more than 500 individuals, an organization not only has to promptly inform all the patients individually but it also must provide "prominent" media outlets within the State or jurisdiction of the breach. That will probably require a press release put out within a reasonable amount of time—no more than 60 days after you detect the breach.

CALCULATING THE COST OF SECURITY

How much will it cost to create an airtight security system that will prevent PHI from being exposed? There is no such thing. No matter how much you invest, you cannot guarantee complete protection to your records. Fortunately, government regulators do not expect it. They expect organizations to take reasonable measures to prevent a breach and to report data exposure should it occur. Data encryption is one such preventive measure.

Encryption, which essentially makes electronic information unreadable by converting it into gibberish until it is unlocked with an encryption key, should be installed on any laptop or other mobile device containing PHI, personally identifiable information, as well as a variety of other types of sensitive data. There are numerous ways to accomplish that, depending on your resources, the skill set of the person who handles your IT operations and your budget.

If a small practice has only a shoestring budget for information technology and there is a consultant or someone on staff with the technical know-how, it is possible to encrypt data on Windows computers by turning on Bitlocker, a build-in encryption tool—assuming you have the correct Windows operating system. Apple computers have a similar tool, called FileVault2.

As you would expect, a more sophisticated encryption system will cost more. You can pay between $250,000 and $500,000 for an enterprise encryption system [11].

The Ponemon Institute has estimated that the average cost of installing full hard disc encryption on a laptop or desktop computer in the United States will run $235 per year. But it also estimated that you are likely to save $4650 as a result of not having your data exposed with said encryption. Put another way, the Ponemon research, which surveyed over 1300 individuals in IT and IT security in the United States, Great Britain, Germany, and Japan, concluded that the benefits of full-disk encryption "exceeded cost in all four countries by a factor ranging from 4 to 20." The study looked at costs in several industries and broke done the results industry by industry. Finance and health care had the highest costs, $388 and $363, respectively [12].

Unless you have an IT professional on staff or an employee with extensive knowledge of health-care IT, you may need to bring in third-party experts to implement many of the other security features needed to be compliant with HIPAA regulations. For the sake of our discussion on the cost of security, you can estimate that it will cost between $50 and $100 an hour for someone to do basic computer and network work; if you want to bring in a security specialist, expect to spend $150–$250 per hour [13].

One security and compliance vendor recently estimated that a small provider would have to pay between $4000 and $12,000 to comply with

HIPAA rules [14]. Decision makers also have to factor in the cost of firewalls, antispyware, and antimalware software. McAfee, for instance, charges about $22–$25 per license for a software package that will cover 250 or fewer devices. Another approach to PHI security is to hire an HIPAA auditing firm to analyze your weaknesses and strengths. In some respects, it is like asking the OCR to come in *before* a breach occurs to investigate where one is likely to happen. These companies review your existing safeguards, do their own risk assessment, and create a risk management plan. You can expect to spend up to 3 months with the auditor and spend at least $40,000 [15].

HOW DO THE HIPAA REGULATIONS FIT IN?

Once a hospital or medical practice is convinced that strong security measures are not only good for their patients but also make sense from a financial perspective, the next logical step is to gain a basic understanding of relevant regulations and how to comply with them.

HHS defines PHI as individually identifiable health information that is transmitted or maintained in any form or medium by a "covered entity" or its business associate (BA) [16]. HHS also defines individually identifiable health information, which it says is health information, including demographics, which "relates to a person's physical or mental health or provision of or payment for health care" and that identifies the individual. The government provides a list of specific elements that are considered part of PHI, including a patient's name, geographic details such as his or her street address, city, state, and zip code. It also includes several relevant dates, such as the patient's date of birth, when they were admitted to or discharged from the hospital. Other patient identifiers that HHS considers sensitive enough to protect include telephone numbers, email addresses, social security numbers, biometric identifiers such as a patient's fingerprints and voiceprints, certificate and license numbers, fax numbers, universal resource locators (URL, also referred to as a web address), medical record identifier, health plan member number, the patient's photo, individually identifiable genetic data, even IP address numbers [13,17].

The HIPAA Privacy Rule cites a medical record, laboratory report, or hospital bill as examples of PHI "because each document would contain a patient's name and/or other identifying information associated with the health data content [18]." But on the other hand, a health plan report that only says that the average age of health plan members is 45 years "would not be PHI because that information, although developed by aggregating information from individual plan member records, does not identify any individual plan members and there is no reasonable basis to believe that it could be used to identify an individual."

Finally, the aforementioned HIPAA rule mentions BAs. HHS defines this term: "A person or entity who, on behalf of a covered entity, performs or assists in performance of a function or activity involving the use or disclosure of individually identifiable health information, such as data analysis, claims processing or administration, utilization review, and quality assurance reviews… Business associates are also persons or entities performing legal, actuarial, accounting, consulting, data aggregation, management, administrative, accreditation, or financial services to or for a covered entity where performing those services involves disclosure of individually identifiable health information by the covered entity or another business associate of the covered entity to that person or entity [19]."

So what do these rules require health-care organizations to actually *do* on a daily basis? They are required to provide patients with a copy of their medical records; they can, however, ask them to make that request in writing if they so choose. The provider has up to 30 days to respond to a patient's request for their records but OCR also states that "As a practical matter, individuals might expect, when making a request of a technologically sophisticated covered entity, that their requests could be responded to instantaneously or well before the current required time-frame [20]."

As a general rule, an organization has to provide patients access to what HHS refers to as "designated record sets," which include medical and billing records, a health plan's enrollment, payment, claims adjudication, and case management records. It also includes any information used by an organization to make decisions about the patient. You can deny access to certain types of patient information, including psychotherapy notes, information for use in legal proceedings, certain information held by clinical labs, and some requests made by prisoners.

The Private Rule requires an organization to seek the permission of patients to share their medical information with others, with some exceptions. Clinicians do not have to ask permission when they share medical data with other clinicians for the purpose of treating the patient—but they need to be especially careful that this information arrives at the correct destination, whether it is in paper, oral or electronic format. Although the focus recently has been on the danger of compromising patients' electronic data through external hacking or snooping by internal users, it is easy to forget about less newsworthy risks.

HIPAA regulations also put a premium on written policy statements and staff training. All the technology in the world cannot replace a clear cut set of institutional guidelines and a culture that values patient safety and privacy. However, HHS realizes that the needs and capabilities of health-care organizations vary widely and attempts to take these differences into consideration as it spells out the administrative requirements needed to mitigate the risk of an information leak.

HHS expects providers to appoint a privacy official to develop and implement the organization's privacy policies and procedures, as well as a person in the office to contact in case there are complaints or requests for information.

Equally important is a workforce training program that educates all workflow members on policies and procedures. And the HIPAA regulation makes it clear that the workforce includes not only employees but also volunteers, trainees, and anyone else whose conduct is under the direct control of your organization, whether or not they are paid for their services. The federal regulations also insist that there be a mechanism in place that applies sanctions against workers who violate policies and procedures in the Privacy Rule. In other words, workers need to be held accountable for their actions and realize that there can be serious consequences for ignoring the privacy safeguards put in place.

Two of the most important requirements spelled out in the HPAA Security Rule center around identifying and protecting against reasonably anticipated threats and protecting ePHI from reasonably anticipated, impermissible uses, or disclosures. To accomplish those twin goals requires a detailed risk analysis and a well thought out management plan—two things that many smaller providers tend to overlook.

The Security Rule outlines four steps in the risk analysis process: (1) evaluate the likelihood and the impact of the potential risk to your ePHI, (2) put the necessary security measures in place to address the risks your analysis has detected, (3) document the measures that have implemented and where required the rationale for these measures, and (4) maintain continuous security protections, periodically evaluating their effectiveness.

The Security Rule spells out three categories of safeguards needed to protect PHI: physical, administrative, and technical. The physical and administrative protocols are similar to those required in the Privacy Rule, including workforce training and physically securing devices in place. But the technical safeguards are worth a closer look.

They fall into four broad categories: access controls, audit controls, integrity controls, and transmission security. Access control refers to the technical policies and procedures that allow only authorized persons to access ePHI. Audit controls are a set of hardware, software, and/or procedural mechanisms to record and examine access and other activity in information systems that contain or use ePHI. Integrity controls refer to policies and procedures to ensure that ePHI is not improperly altered or destroyed. It also requires electronic measures be put in place to confirm that ePHI has not been improperly altered or destroyed. And lastly, transmission security means implementing technical security measures that guard against unauthorized access to ePHI that is being transmitted over an electronic network [21].

TAKING THE NECESSARY PREVENTIVE MEASURES

There is a long list of technological tools that can help health-care organizations protect patient information, including encryption, user authentication, mobile device management software, firewalls, and antimalware programs, but an effective preventive strategy requires a more holistic perspective. As Tom Walsh, a veteran security specialist, once explained it, any attempt to strengthen one's defenses requires you look at three important areas of concern: people, processes, and technology [22].

Your weakest link is always people, those troublesome "carbon-based interface units" as Walsh calls them. Several large-scale data breaches have been traced to employees being duped by various phishing scams, for example. During these scams, employees click on an email attachment from a "friend" or other seemingly trustworthy source only to be sent to a website that loads malware onto the individual's computer or network the employee works on. One of the most effective ways to block such intrusions is by educating your physicians, nurses, administrators, BAs, and anyone else who has access to patient data.

The people problem can manifest itself in other ways as well. If you put in place security procedures that make it too difficult for clinicians to provide quality patient care, they are probably going to rebel, or look for workarounds that are less secure. Clinicians may need to compromise and accept a measure of inconvenience to make their online activity more secure, but by the same token, security measures need to meet a reasonable standard, and they need to be understandable to the clinicians and administrative staff who use them. That once again requires well-thought out employee training and a set of sensible policies and procedures that are distributed to everyone in the organization.

Since people are the weakest link in the security chain, let's start with them, and their vulnerability to email and Internet scams that use social engineering. Social engineering essentially taps into many of the normal human personality traits that allow a society to function, traits like the desire to help out persons in need, the quest for recognition or financial gain, everyone's natural curiosity about their neighbors' affairs, just to name a few. Hackers rely on these tendencies to convince potential victims to open infected emails or websites. A report from Trend Micro suggests that more than 90% of cyberattacks begin with such spear phishing emails [23].

Spear phishing refers to the targeted nature of the emails sent to potential victims. They may call you by name, mention your job title, or mention other personal information that email recipients assume can only originate with friends or BAs, or companies that you already have a relationship with. If a hacker has already infected one person's machine and gains access to their address book, he or she can then send phishing emails to those on that list. Since the intended new victim sees a friend's address, they often assume the message is legitimate.

There are so many ingenious ways to create a convincing phishing email that the best approach to preventing being duped is to assume that almost *every* message that arrives in your inbox is a scam until proven otherwise: guilty until proven innocent.

The most important piece of advice you can give staffers is: Do not click on hyperlinks embedded in an email, unless you are absolutely certain it is from a legitimate source. Unfortunately, many people do not believe that they would ever fall for such trickery. "I am too smart to be fooled by social engineering tricks." One way to convince them otherwise is by running a fake phishing scam. Tom Cochran, formerly in charge of White House digital technology, was able to convince his coworkers at Atlanta Media that they were easy prey by sending out a fake phishing email to all the employees at the firm. Within 2 h, he had his proof: "Almost half of the company opened the email, and 58% of those employees clicked the faux malicious link" [24]. These statistics were far more convincing to staffers than a memo mandating that they follow certain precautionary steps. In Cochran's view: "Placing someone in a cyberattack drill is the safest and most effective tactic to build the company's collective security intelligence."

PASSWORDS, POLICIES, AND PROCEDURES

Processes, the second item on the list of broad security issues mentioned previously, brings us to the three Ps: Passwords, policies, and procedures. Passwords fall under the category "user authentication," which is tech speak for the process of verifying that the person signing on to a computer system really *is* the person he or she claims to be. To balance strong security against user convenience, passwords need to be hard to guess but not too hard for staffers to remember.

The password policy at Beth Israel Deaconess Medical Center (BIDMC) is worth emulating. Passwords need to be at least eight characters long and must consist of four types of characters: uppercase letters, lower case letters, numerals, and special characters such as @ or #. Skeptical employees may feel this is overkill but they are dead wrong! There are several password cracking software programs available that allow hackers to scan millions of common passwords *per second* to locate yours. These tools typically include virtually every word in the dictionary, as well as common phrases from popular and classical literature.

Of course, asking staffers to come up with passwords that meet all the aforementioned criteria will get pushback because they are harder to remember than those that rely on a pet's name or a favorite TV character. One option is to suggest they think of a passphrase or short memorable sentence and then shorten it by using initials. So for instance, "I live at 322 Grand Avenue in Brooklyn" can become Il@322GAiB (Notice that the password has the same upper and lower case letters that exist in the sentence, which makes it easier to remember). It is probably best not to use one's actual street address in the passphrase, however, since intruders may have access to that information. And whatever password is used, never *ever* write it on a sticky note and paste it to a nearby workstation. That is one of the first violations HIPAA auditors will look for when they do a site inspection.

It is also advisable to choose a user authentication system that forces employees to follow the aforementioned guidelines on password strength; in other words, a password generation technology that rejects the creation of new passwords that are not at least eight characters long, uses a combination of lower and uppercase letters, and requires at least one special character. To reduce the likelihood of unauthorized persons gaining access to your network by means of a password cracking program, it is also wise to lock out users who fail to enter the correct code after 10 attempts.

One of the weaknesses in US federal agencies has been user authentication. After a recent data leak, some agencies have only now begun to use a two-factor approach to verify all of their authorized users, which has been considered a basic precaution in the minds of many security specialists for years. In single factor authentication, once you enter your user name and password, you are allowed entry into the system. The two-factor approach requires a second step to harden security. It may require you to answer a few questions, for example, what was your best friend's last name or your grandmother's maiden name. Or it may require swiping an ID card or a token, or a biometric scan of your fingerprint or retina.

If the second authentication procedure is too annoying, users will rebel, which is why typing in answers to one or two questions is more palatable than carrying a smart card around, and less expensive than biometric scans. However, if an organization already requires employees to wear their ID badges, adding the technology that allows it to function as the second leg in the authentication procedure is much more palatable.

HIPAA states that a health-care organization must "verify the identity of a person requesting protected health information and the authority of any such person to have access to protected health information… if the identity or any such authority of such person is not known to the covered entity" [25]. Although that regulation does not specify two-factor method, it is worth serious consideration.

TECHNOLOGICAL SOLUTIONS

The third component in a holistic prevention strategy, technology, includes a long list of tools that may prove challenging to decision makers who did not get their degrees in computer science or work their way up the ladder learning to implement and manage these tools. The list includes firewalls, antiviral and antispyware programs, intrusion prevention systems, encryption, user authentication protocols, and audit logs. A deep understanding of these safeguards is beyond the scope of this book but our aim here is to provide enough details to allow C-suite executives and other decision makers to offer direction to those with more technical expertise.

Encryption is a way to disguise text or other information so that it is not recognizable to others. This means converting characters in the message into gibberish so that they cannot be read by unauthorized persons, and then having a way to decode or "decrypt" the message so that it can be read by authorized persons. To oversimplify the process, it involves turning the letters a, b, and c into x, j, and q (Technically speaking, this process is called a substitution cypher, which is encoding, not encryption). Unfortunately, in today's world, simple substitution of one letter for other is far too easy for hackers to decode, so modern cryptographers use sophisticated algorithms and protect them with encryption and decryption keys that prevent others from deciphering the patient data that are supposed to remain confidential. The original patient information is referred to as plaintext and the encoded information is ciphertext. The algorithm converts the plaintext to ciphertext, which is based on a set of rules that tells the computer how to translate between the plaintext and the encrypted messages. For example, if we were to look at a simple substitution cipher, the rule might call for the conversion of every letter to three letters later in the alphabet, thus substituting every "a" character to a "d," "b" to an "e," and so on.

Modern cryptographic algorithms are typically based around the use of keys. The encryption key serves as the mechanism to instruct the computer how to translate the ciphertext back into a plaintext message. These keys usually come in two forms. The first system, called symmetric encryption, requires a single key to encrypt and decrypt the message. A second approach, referred to as a public key or asymmetric cryptography, makes use of a publically available key for encryption and a separate private key for decrypting. The latter is considered more secure but is a longer process and requires more processing power. Additional safeguards that will help strengthen the walls around an organization's PHI include a virtual private network (VPN)— which is illustrated in Fig. 9.1—and mobile device management software, which are discussed in more detail in *Protecting Patient Information* [26].

APPLYING SECURITY PRINCIPLES TO PRECISION MEDICINE INITIATIVE

While the privacy and security principles that apply to hospitals and medical practices in general are relevant to the PMI, those principles are only the starting point for a massive, nationwide program that will collect data on a million or more Americans. The US government has developed several guidelines to help keep the PMI project safe. They include *PMI: Data Security Policy Principles and Framework* from the White House, and *PMI Data Security Principles Implementation Guide*, developed by the HHS Office of the National Coordinator for Health Information Technology (ONC).

The implementation guide was designed to help organizations that use patient data for research, including those that are participating in the PMI

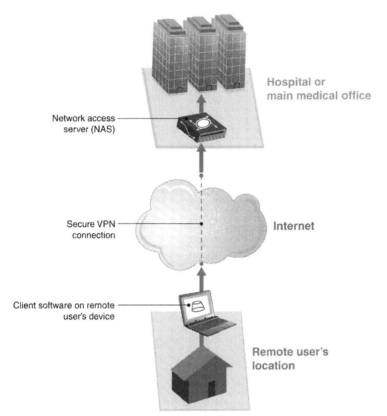

Figure 9.1 The path taken by patient information as it moves through a virtual private network.

or the National Institutes of Health All of UsSM Research Program; organizations using PMI data outside these federal programs, and contractors working with PMI data for research purposes. Taking a pragmatic approach, the guide walks readers through a case scenario in which a fictional Chief Information Security Officer conducts a risk analysis of a data system and then outlines the steps that said CISO follows to address the problems identified in the risk assessment.

A CASE SCENARIO REQUIRING SECURITY PROTOCOLS

The ONC guide sets up the illustrative scenario with a diagram and summary of what a fictional researcher might try to accomplish. In this scenario, the investigator works for a research organization that has partnered with an academic medical center. As Fig. 9.2 shows, she uses a laptop to remotely connect with the University's research information system (RIS),

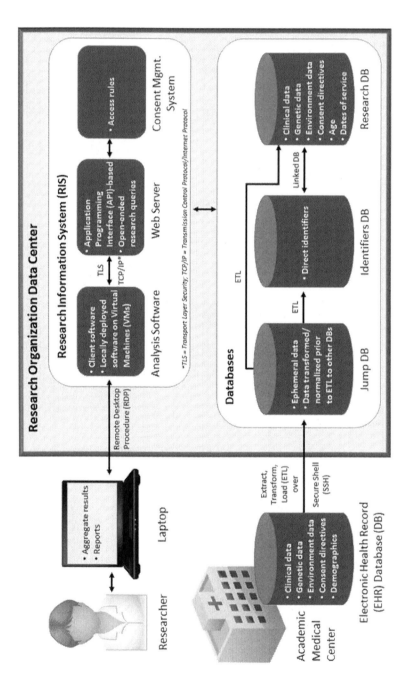

Figure 9.2 The diagram illustrates the path that protected health information can take through a research data center to keep it secure and private. *(From HHS Office of National Coordinator for Health Information Technology. Precision medicine initiative (PMI) data security principles implementation guide.)*

which itself is part of a larger data center. The RIS consists of three "buckets": the analysis software, a web server, and a consent management system, which are all interconnected.

The data center also houses three databases that the medical center has access to. It too consists of three components: a jump database, an identifiers database, and a research database that contains all the vital information the investigator needs to conduct her analysis. That includes clinical, genetic, environmental data, consent directives, patient demographics, etc.

The academic medical center is the original source of all this research data, which it has stored in an electronic health records system. The data are sent to the data center through a process called extract, transform, and load and are transmitted through a secure shell tunnel.

Before the CISO even steps into the picture, it is clear that the IT team has already put in place several privacy and security measures to keep patient records safe. To begin with, the outside researcher is not given direct access to the hospital's EHR system; she only has access to that data in the research database within the Data center. She gains access to it through a web server. In addition, there are certain rules to limit the kind of data she can see, rules that are controlled by the consent management system. Other precautions include the transport layer security that exists between the client software and the web server.

In the ONC case study, it is understood that the contractor and the researcher doing the data analysis are not bound by HIPAA rules for this project. It is also assumed that the risk analysis that the CISO is performing only applies to this one project—not to the entire enterprise.

Once the CISO identifies any security risks that may exist within the PMI-related research project, she will rank the severity of those risks using a tool called the Common Vulnerability Scoring System version 3. It ranks the potential dangers in three dimensions: a base score, temporal score, and environmental score. ONC explains the scoring protocol:

> *The base score captures the characteristics of the vulnerability that are constant over time and across environments. This score is modified by the temporal and environmental score. The temporal score captures how risks may change over time. For example, vulnerabilities may initially be hard to exploit and thus have a low risk score, but later may be included in an easy-to-use exploit software which raises the score. The environmental score captures characteristics that are unique to the user's environment.*

The process of performing a risk analysis of a PMI research project has many of the same features as a risk analysis for a medical practice or hospital, which we discussed earlier in this chapter. And the preventive measures required to address any vulnerabilities are similar. Once a health-care organization has found the problems, it can choose several strategies to address the risks that have been identified. The ONC Implementation Guide summarizes these measures as follows:

- Implementing controls that reduce the risk. For example, antivirus and antimalware solutions directly reduce the risk of malicious software.
- Implementing compensating controls. For example, patching legacy systems is the ideal, but legacy systems may not have security patches available. Isolating legacy systems on a separate network without access to the Internet is a compensating control that reduces the risk that unpatched systems will be breached.
- Ensuring risks are addressed in contracts. Risks may arise because of relationships between the research contractor and consultants, suppliers, or other research institutions. All of these relationships are documented in contracts. The relevant contracts need to specifically address privacy and security issues, including financial responsibility for privacy and security through liability terms and indemnification clauses.
- Purchasing cyber insurance to cover the risk. Cyber insurance can blunt the impact that security risks have on a business. Organizations should carefully review cyber insurance policies to understand the scope of the policies and ensure that identified risks are actually covered.
- Destroying the data. Data that have been properly destroyed are no longer at risk of being improperly accessed or modified. Organizations should regularly review their document and data retention policies to ensure that old data are not being held for longer than needed by the business. Organizations may also need to consult state law, as data retention policies often vary by state.
- Accepting the risk. Not all risks can be eliminated—organizations should determine the level of risk with which they are comfortable.

REFERENCES

[1] Conn J. Advocate data breach highlights lack of encryption, a widespread issue. Modern Healthcare; 2013. http://www.modernhealthcare.com/article/20130830/NEWS/308309953.

[2] Kam R, Ponemon L. Why healthcare data breaches are a C-suite concern. Forbes; 2012. http://www.forbes.com/sites/ciocentral/2012/12/07/why-healthcare-data-breaches-are-a-c-suiteconcern/.

[3] McCann E. Healthcare's slack security costs $1.6B. Healthcare IT News; 2014. http://www.healthcareitnews.com/news/healthcares-slack-security-costs-16b.

[4] U.S. Department of Health $ Human Services. HHS announces first HIPAA breach settlement involving less than 500 patients, Hospice of North Idaho settles HIPAA security case for $50,000. 2013. http://www.hhs.gov/ocr/privacy/hipaa/enforcement/examples/honi-agreement.html.

[5] Fulford M. OCR audits: don't fall victim to past mistakes. 2014. http://www. informationweek.com/healthcare/security-and-privacy/ocr-audits-dont-fall-victim-to-past-mistakes/a/d-id/1317645.

[6] Privacy Rights Clearinghouse. Fact sheet 8a: health privacy: HIPAA basics, How does HHS determine a penalty for a violation? https://www.privacyrights.org/content/health-privacyhipaa-basics#hhs-determine-penalties.

[7] U.S. Department of Health & Human Services. Massachusetts general hospital settles potential HIPAA violations. http://www.hhs.gov/ocr/privacy/hipaa/enforcement/examples/massgeneralra.html.

[8] Ornstein C. Policing patient privacy: fines remain rare even as health data breaches multiply. ProPublica; 2015. http://www.propublica.org/article/fines-remain-rare-even-as-health-data-breachesmultiply.

[9] Stapleton T. Data breach cost: risks, costs, and mitigation strategies for data breaches. Zurich American Insurance Corporation; 2012. http://www.zurichna.com/internet/zna/sitecollectiondocuments/en/products/securityandprivacy/data%20breach%20costs%20wp%20part%201%20%28risks,%20costs%20and%20mitigation%20strategies%29.pdf.

[10] Munro D. Assessing the financial impact of 4.5 million stolen health records. Forbes; 2014. http://www.forbes.com/sites/danmunro/2014/08/24/assessing-the-financial-impact-of-4-5-millionstolen-health-records/.

[11] McMillan M, Cerrato P. Healthcare data breaches cost more than you think. Information-Week Healthcare Report; 2014.

[12] Information Week Network Computing. Calculating the cost of full disk encryption. 2012. http://www.networkcomputing.com/careers-and-certifications/calculating-the-cost-of-full-diskencryption/d/d-id/1233859.

[13] Herold R, Beaver K. The practical guide to HIPAA privacy and security compliance. 2nd ed. Boca Raton (FL): CRC Press; 2015. 16-L 17.

[14] Ferran T. How much does HIPAA compliance cost? http://blog.securitymetrics.com/2015/04/how-much-does-hipaa-cost.html.

[15] Ferran T. How much does a HIPAA Risk management plan cost? Security Metrics Blog; 2015. http://blog.securitymetrics.com/2015/01/how-much-does-hipaa-risk-management-cost.html.

[16] Department of Health and Human Services. Protected health information. http://www.hhs.gov/ocr/privacy/hipaa/understanding/training/udmn.pdf.

[17] HHS NIH 2. Department of Health and Human Services National Institutes of Health, How can covered entities use and disclose protected health information for research and comply with the privacy rule? http://privacyruleandresearch.nih.gov/pr_08.asp.

[18] HHS.gov. Health information privacy, guidance regarding methods for de-identification of protected health information in accordance with the Health Insurance Portability and Accountability Act (HIPAA) privacy rule. http://www.hhs.gov/ocr/privacy/hipaa/understanding/coveredentities/De-identification/guidance.html.

[19] Department of Health and Human Services National Institutes of Health. To whom does the privacy rule apply and whom will it affect? http://privacyruleandresearch.nih.gov/pr_06.asp.

[20] HIPAA Office of Civil Rights (OCR). The HIPAA privacy rule's right of access and health information technology. http://www.hhs.gov/ocr/privacy/hipaa/understanding/special/healthit/eaccess.pdf.

[21] HHS.gov. Summary of HIPAA security rule. http://www.hhs.gov/ocr/privacy/hipaa/understanding/srsummary.html.

[22] Walsh T. Risk management and strategic planning. In: Herzig TW, editor. Information security in healthcare. Chicago: Healthcare Information and Management Systems Society; 2010. p. 18.

[23] Savvas A. 91% of cyberattacks begin with spear phishing email. Techworld; 2012. http://www.techworld.com/news/security/91-of-cyberattacks-begin-with-spear-phishing-email-3413574/.

[24] Cochran T. Why I phished my own company. Harvard Business Review; 2013. https://hbr.org/2013/06/why-i-phished-my-own-company/.

[25] Cornell University Law School. 45 CFR 164.514-other requirements relating to uses and disclosures of protected health information. https://www.law.cornell.edu/cfr/text/45/164.514.

[26] Cerrato P. Protecting patient information: a decision-maker's guide to risk, prevention, and damage control. New York: Syngress/Elsevier; 2016.

CHAPTER TEN

Patient and Consumer Engagement

History has a way of revealing what resistant minds refuse to see. A case in point: In 1977, Donald Ardell wrote a book for the consumer press entitled *High Level Wellness*. It was one of the early attempts to promote holistic health. At the time, the concept was rather radical because much of patient care was still focused on finding single causes for single diseases. And for good reason: Specific bacterial strains had been identified as causative agents for specific diseases such as pneumonia and meningitis; specific environmental toxins such as lead were linked to neurocognitive disorders; deficiencies of individual vitamins and minerals had been shown to cause specific nutritional disorders such as pellagra, osteomalacia, and beri beri. The logic seemed sound.

But Ardell proposed that poor health was the result of numerous interacting factors, including microbes, psychosocial issues, nutritional imbalances, lack of adequate physical activity, exposure to environmental pollutants, and more. That mindset met resistance from the medical community because it challenged conventional wisdom at the time—and because Ardell was unable to support his thesis with much scientific evidence. The book has very few references to the biomedical research literature, in part because there was not much evidence available at the time. But the scenario has changed significantly in the past 40 years. Many of the causative factors Ardell wrote about have now been shown to contribute in varying degrees to several diseases. This new evidence is slowly ushering in an era of *data-driven* holistic medical care that is enabling clinicians to identify the interconnected causes of each individual's health problems. What remains to be accomplished now is to gain the public's trust and to engage them in the quest for high-level wellness.

The journey to consumer and patient engagement begins with another issue that Ardell discussed: self-responsibility. It is one of the most critical ingredients required for the public to achieve optimal health and personalized medical care. Unfortunately, many Americans take better care of their

Realizing the Promise of Precision Medicine
ISBN 978-0-12-811635-7
http://dx.doi.org/10.1016/B978-0-12-811635-7.00010-5

183

cars and wardrobes than their bodies. If they develop a health problem, they expect the doctor to fix it, much like they expect their car mechanic to fix a flat tire or replace a faulty catalytic converter.

Even if clinicians had an in-depth understanding of all the myriad contributing cause of disease and could measure in infinite detail the extent to which genes, stress, nutrition, pollution, climate, and specific bacterial and viral strains were contributing to each patient's health problems, this understanding would not be enough if the public remained passive participants in care. And in today's quick-fix, fast temporary relief society, convincing patients to become active participants in their own care is a hard sell. We nonetheless need to sell it.

ENGAGING RESPONSIBLE PATIENTS

How can clinicians engage patients in their own care and spark a sense of self-responsibility? Admittedly the opportunities to accomplish this dual feat are limited, given the fact that the average office visit lasts about 10 min—and given the fact that like any relationship, the doctor–patient relationship requires commitments from both parties. But if patients are willing to meet us halfway, there is much that can be accomplished.

The first step is for clinicians to ask themselves: Do I need better communication skills? Adrienne Boissey, MD, MA, Medical Director of the Center of Excellence in Healthcare Communication at the Cleveland Clinic, addresses this sensitive topic in the course she oversees: "Adults choose to learn, or not. But if you don't have an interest in learning or don't think you have any blind spots, you absolutely won't engage in the learning process." This blind spot is sometimes referred to as the Dunning Kruger effect, which is a cognitive bias that essentially causes one to be ignorant of one's own ignorance.

With expertise in evidence-based learning theory, Boissey explains the same phenomenon in different terms, calling it "unconscious incompetence." One of the ways that physicians can learn to become conscious of their weaknesses as communicators is to have their shortcomings presented to them through the comments of patients who they have worked with.

Boissey states: "Putting patient comments about how they felt when you communicated with them, about how effective your language was, and how that made them feel—putting that back at the physicians, and showing them, 'this is how patients felt during their interaction with you'—I think is a very powerful way of driving that interest and change." [1]

The 8-hour Cleveland Clinic course that addresses these issues has been shown to produce measurable results, generating a statistically significant impact of patient experience, empathy, and burnout. Those results should not be surprising in light of the fact that there is 30 years of research to show that good communications prevents malpractice, improves patient safety, and physician satisfaction.

The Cleveland Clinic course goes beyond the usual discussion about the value of being a good listener and other fundamentals, and seeks to be more transformative, to change attitudes and beliefs, and to emphasize the need for relationships. For example, it asks: "What would your language sound like if you were trying to build a relationship with someone, as opposed to just trying to get them to take their medicine?...Can you think about what your role or responsibility in that would be, as a physician?"

Of course for some physicians, the path to good communication will be difficult. A recent installment in Gary Trudeau's comic Strip Doonesbury makes this point. The scene opens with Gary putting his shirt back on after his physical, with the doctor, an older gray-haired gentleman, looking at his computer tablet and entering data.

Gary asks: "Say Harry, I expect you saw that new study about women doctors…"

No response from the doctor as he keeps tapping his computer screen.

Gary: "…whose patient survival rate is higher than that of male doctors? Apparently women doctors are more likely to follow treatment guidelines and provide preventative care. They also communicate better. To be honest, Harry, all that sort of resonated with me."

No response from the doctor, who turns back to his tablet, uttering one word: "Hunh."

No doubt many patients can relate to this scenario. Numerous articles have been written about the importance of not becoming too obsessed with data entry, the need to look at patients rather than computer screens, and the benefits of positioning the office computer so that one can easily look at the patient between data entries. To solve the problem, some clinicians have even hired scribes, who are specifically hired to handle note taking during patient exams and interviews.

While it may seem too obvious to discuss the value of *talking* to patients to get them more engaged, it nonetheless needs to be mentioned in today's fast-paced world. A good example is the conversation that should take place between physicians and patients with chronic kidney disease. It is usually best for a nephrologist to talk to these patients about the importance of having a vascular surgeon inserting a fistula or graft months before they begin hemodialysis. Unfortunately, in the rush of everyday medical practice, clinicians too often do not take the time for this conversation. The result is that many patients start dialysis with a catheter instead, which increases the likelihood of infections and other complications. It has been estimated that the needed conversation costs less than $200; the cost of complications increases the cost of care by more than $20,000 [2].

THE RETURN ON INVESTMENT

Studies have also demonstrated that when clinicians spend time addressing patients' concerns and explaining how to manage chronic conditions in general, treatment adherence improves and fewer expensive complications develop. This has been documented in orthopedic practices, for instance, in which 30–60 min educational sessions conducted by nurses led to shorter hospital stays, fewer complications, and high rates of discharge to home rather than postacute care facilities [2].

To document the value of patient engagement and patient activation, researchers have developed a validated tool called the Patient Activation Measure (PAM). By means of a series of questions and answers, PAM arrives at a score of 0–100 that estimates how responsible a person is in terms of managing his/her own health. Judith Hibbard at the University of Oregon and Jessica Greene from George Washington University School of Nursing explain [3]:

> The score incorporates responses to thirteen statements about beliefs, confidence in managing health-related tasks, and self-assessed knowledge. Examples include the following: "I am confident that I can tell whether I need to go to the doctor or whether I can take care of a health problem myself"; "I know what treatments are available for my health problems"; and "I am confident that I can tell a doctor my concerns, even when he or she does not ask." Responses are degrees of agreement or disagreement. The measure has been proved to be reliable and valid across different languages, cultures, demographic groups, and health statuses.

PAM reliably predicts whether patients will have regular check-ups, take preventative measures, have immunizations, eat better, and get regular exercise. Similarly activated patients have been found to have a body mass index, hemoglobin A1c (HbA1c), blood pressure, and cholesterol in the normal range [3]. Hospitalization and (emergency department) ED costs are also lower for highly activated patients, when compared to those who are less motivated. And less-motivated patients are almost twice as likely to be readmitted to the hospital 30 days after discharge, when compared to more active patients [4].

Patient engagement has been called the blockbuster drug of the 21st century. It certainly has the potential to be so. But for this drug to be truly effective, it must be "metabolized" by a willing recipient. In fact, for the personalized medicine model to work, it demands patients have the skills to fully participate in their care. Several clinical trials have been conducted to determine what skills are needed and whether or not they can be taught to patients.

Simmons and associates have analyzed 10 randomized controlled clinical trials in which various educational and motivational interventions were used to boost patient engagement and activation. The five studies that looked at diabetes management found that the interventions all improved patient engagement. One investigation, for example, involved 12 months of internet-based computer-assisted self-monitoring plus social support; it found that the interventions improved nutritional habits and physical activity levels, as well as HbA1c and several other biological parameters. In short, Simmons et al. demonstrated "the positive effects of disease management interventions on increasing patient engagement and improving health outcomes." [5] The analysis was convincing enough to prompt the researchers to conclude that an individual's engagement commitment should be measured as a risk factor during a patient's workup. The implications of their research are straightforward: Scoring poorly on a PAM questionnaire might be as threatening as scoring poorly for other well-documented risk factors, including obesity, smoking, heavy alcohol use, elevated serum cholesterol, and hypertension.

PATIENT PORTALS ARE NOT ENOUGH

Many health-care organizations have created online patient portals to meet the requirements outlined in the Centers for Medicare & Medicaid Services (CMS) Electronic Health Record (EHR) Incentive Program or Meaningful Use Program. That has certainly been an important step toward engaging patients in their own care, but the Meaningful Use program is better described as a "Minimum Use" program since it proscribed the bare minimum needed to stimulate patient activation. Many portals are little more than a depository that houses patients' laboratory results and the practice's office policies.

There are several ways to enhance the patient portal, enhancements that make them more patient-centric. There are several role models worth emulating, including Kaiser Permanente and the Cleveland Clinic. But even providers with very limited resources can still wow their patients with a little creativity, a talented writer, and the help of a Web designer [6].

But even the best patient portals are no substitute for a mindset that encourages shared decision making (SDM). In the not-too-distant past, sharing treatment decisions with patients were considered unprofessional since the public was "too uninformed" to know what was best for them. Such paternalism is well-represented in an excerpt from a 19th century article published in the *Boston Medical and Surgical Journal*, which eventually came to be called the *New England Journal of Medicine*:

> *The question is often asked, why physicians do not write…prescriptions in English. The answer is obvious — that if they did, the patient would often be less benefited than he now is. There are very few minds which have sufficient firmness, during the continuance of disease, to reason calmly on the probable effects of remedies, and to compare their wonted action…with the indication to be fulfilled in the particular case…. The only state in which the mind can rest…during severe illness, is that of implicit reliance in the skill of the physician, and an entire acquiescence in the course adopted, without the slightest question or argument.*
>
> **Schiff et al. [7]**

Most 21st century patients would be enraged with this type of conduct from their provider. Most are better informed today and many want to have a say in how they are managed. But there is another side to the issue. It's summed up succinctly on coffee mugs found on many physicians' desks: "Please don't confuse your Google search with my medical degree." Clearly, there needs to be a balance between two extremes. Letting activated patient dictate the treatment protocol can be just as harmful as taking a "Father Knows Best" approach to patient care.

SHARED DECISION MAKING

The term shared decision making has a long history, originally coined by a Presidential Commission in 1982. In 2000, the prestigious Institute of Medicine published a report entitled *Crossing the Quality Chasm*, the intent of which was to encourage a redesign of the health-care system. It included a call to action to make SDM a reality and to give patients more control over their care. Among the 10 rules of redesign spelled out in the report [8]:

- The patient is the source of control.
- Knowledge is shared and information flows freely.
- Transparency is necessary.

The report went out to state: "Patients should be given the necessary information and opportunity to exercise the degree of control they choose over health care decisions that affect them. The system should be able to accommodate differences in patient preferences and encourage shared decision making."

Health care in the United States has yet to fulfill that mandate, but there are indications we are moving in that direction. CMS recently decided to put its weight behind the concept by launching a "Shared Decision Making Model" to help Medicare patients have more input into treatment options [9]. According to CMS, "Although it has been 30 years since the Commission urged the adoption of shared decision making, beneficiary preferences and values about medical treatment choices are still routinely left out of important discussions between practitioner and beneficiary." The agency goes on to explain: "Shared decision making can ensure that treatment decisions, for the many medical conditions that do not have one clearly superior course of treatment, better align with beneficiaries' preferences and values. One facet of shared decision making is the use of patient decision aids (PDAs)— tools that present information about common medical choices. A 2014 systematic review of Patient Decision Aids (PDAs) concluded that using PDAs can help patients gain knowledge; have a more accurate understanding of risks, harms, and benefits; feel less conflicted about decisions; and rate themselves as less passive and less often undecided. These tools do not supplant physician-beneficiary conversations about treatment options; instead, they supplement and/or encourage it by better preparing beneficiaries to engage in those conversations."

Among the reasons that SDM has not been fully incorporated into routine clinical practice: overworked physicians, insufficient practitioner training, inadequate clinical information systems, lack of consistent methods to measure that SDM is taking place, and uncertainty as to whether, or how, to promote change and invest in the time, tools, and training required to achieve meaningful SDM. To overcome some of these obstacles, CMS is testing a four-step process for participating Accountable Care Organizations (ACOs), consisting of:

- Identifying SDM eligible beneficiaries,
- Distributing the PDA to eligible beneficiaries
- Furnishing the SDM Service
- SDM tracking and reporting

The goal is to recruit 50 ACOs nationwide in the project and to compare their performance to 50 control ACOs. Only ACOs participating in the Medicare Shared Savings Program or Next Generation ACO Model are eligible for selection to participate in the SDM Model. More details on the project are available on the CMS website.

One aspect of this pilot program gets at the heart of the balanced approach we spoke of earlier. The program aims to encourage patients and clinicians to share decision-making "for the many medical conditions that do not have one *clearly superior* course of treatment." When a patient comes into the office or clinic insisting that they be given an antibiotic for a viral upper respiratory infection, there is no ambiguity involved in the decision to withhold the medication. It's contrary to all the scientific evidence. Similarly a parent that insists childhood vaccinations should be avoided because they cause autism or a host of other serious complications should not be encouraged under the guise of SDM; the correct course of treatment is clearly indicated by the overwhelming evidence supporting the protective effects of vaccines.

But there are numerous scenarios in which the best treatment option is not so obvious. And these are the situations in which patients have the right to participate in the final choice. And that choice needs to take into consideration not only the scientific evidence for and against a specific therapy but also the patient's financial status, as well as their social and family situation. In light of the increased financial burden that most patients must face as insurance policies demand much larger out-of-pocket costs, sharing the decision is ethically justified. Unfortunately, clinicians have almost no experience counseling patients on to how to weigh the economic issues because they have been shielded from the payment process. Few patients pay their practitioners directly for their services and many hospitals cannot even give patients an exact price quote for specific procedures or services.

There is one other valid reason to encourage SDM: Sometimes the patient is right, i.e., they actually have a better understanding of their condition and how best to manage it than their provider. The experience of Michael Snyder, PhD, a professor at Stanford University School of Medicine, drives home this point. When he sequenced his entire genome, he saw risk markers that convinced him that had diabetes. But every clinician he spoke to said it was impossible. He was slender, had no family history, and his blood glucose levels were normal. Once he was finally able to have a 3-h fasting glucose test performed, however, and an HbA1c test, his suspicions were confirmed [10].

PATIENT-GENERATED DATA AND MOBILE ENGAGEMENT

As we brought out in Chapter 5, there are numerous mobile apps that can help patients become more fully involved in the own care. One way they can help foster patient participation is by collecting health data that normally is not available to clinicians during an office visit. And since most of a person's life is spent outside the confines of a medical office visit, there is a great deal of useful data worth collecting. That can include fluctuations in weight for patients with congestive heart failure, as well as blood pressure and blood glucose readings, to name a few. What also makes this kind of information valuable is the fact that it's longitudinal rather than cross-sectional. A lab test for fasting blood glucose measures one moment in a patient's life, while continuous readings in the real world, coupled with a mobile app that tracks dietary patterns, stress levels, and exercise over many weeks, can be invaluable in detecting obscure causes of hypo- or hyperglycemia.

Apple's HealthKit is also letting patients share their data with clinicians. HealthKit was an enabler that led Beth Israel Deaconess Medical Center (BIDMC) to create BIDMC@Home, an iPhone and iPad app that uploads the internet of things (blood pressure cuff, glucometer, scale, activity, sleep data, etc.) to its electronic health record. CareKit, also an Apple invention, takes us one step further on the wellness-focused journey. Apple's middleware (HealthKit, Research kit, Carekit) has enabled BIDMC to connect devices in patients' homes and enabled the medical center to collect answers to clinician-generated questionnaires with dashboarding of the subjective and objective-combined results [11].

Several other initiatives have demonstrated the value of patient-generated data. One such program helps clinicians monitor chemotherapy-related toxicity by collecting patient self-reports. Another accepts input from HIV-positive patients to help providers evaluate symptoms. A third involves patient reports of pain and functional problems over time to help clinicians assess the effect of physical therapy and medication and to help determine if surgery is needed in patients who may require knee replacement. There are also programs that not only track signs and symptoms but also collect outcomes data from patients. The Orchestra Project, for instance, has a tool that lets children with inflammatory bowel disorder monitor outcomes as they relate to lifestyle and treatment changes [12].

Lavallee et al. point out, however, that there are obstacles that slow down many efforts to engage patients. Among the logistical issues involved in collecting this kind of data are workflow disruptions and the need for extra time on the part of staffers to gather and interpret the meaning of data as it comes in. Among the technological issues are EHR systems that are not capable of assimilating patient-generated data, including diet and physical activity levels.

FINAL THOUGHTS ON THE PROMISE OF PRECISION/ PERSONALIZED MEDICINE

Individuality is an easy concept to grasp but a stubbornly difficult one to apply in patient care. Although clinicians must address the needs of individual patients one by one, they are expected to apply evidence-based medicine and the results of controlled randomized trials in routine practice. Unfortunately, most of that research only tells us how to diagnose and treat large groups of patients, namely the cohorts enrolled in these clinical trials. As the previous chapters emphasize, this one size fits all approach to patient care has its limitations and needs to be supplemented—if not replaced— with an approach that fits each individual's unique biochemical, physiological, and genetic makeup.

As we enter the era of precision/personalized medicine, even the concept of cure needs to be rethought. In 1969, John Halamka's grandfather died of gastric cancer. That year, an advertisement appeared in the Washington Post noting that the "cure for cancer" was right around the corner if we just provided additional funding. But by 2017, our understanding of pathology suggests there will never be a single cure for cancer. Disease is a function of the genotype, the phenotype, the microbiome, the exposome (where you live, the work you do), and diet. Any cure will need to take all of these into account.

Further, every patient has his/her own care preferences and privacy preference. What is your tolerance for risk? How important is longevity versus quality of life? How much data are you willing to share for what purpose?

This book has raised a number of the issues in precision medicine, including the need to collect and share patient data among care teams, patients, and families; the need for genomic interpretation services; the imperative to provide decision support in electronic health records that combines medical data, sequencing data, and patient care preferences.

Some may think that precision medicine is a distant future. This book has provided the evidence to demonstrate that the technology is already here, it just needs to be more evenly distributed and adopted.

Our medical records were on paper when we were in our 20s. Our children's records are now digitalized. The management of Kathy Halamka's cancer, which we discussed in the Prologue, was informed by big data analytics done on thousands of similar patients who were treated in the past. Her experience suggests that evidence-based medicine, with its overemphasis on large-scale randomized clinical trials, needs to encompass the observations gleaned from electronic health records, mobile medical apps, and numerous other sources.

That does not imply that the medical care of the future will be fully automated and computer-dependent. Although "Watson" may be able to read the *New England Journal of Medicine*, it cannot replace human clinicians. In reality, only a partnership between humanity and technology will lead to better cures. Every person is unique. Wellness means something different in every cultural context. Using data, evidence, and experience we can tailor care to optimize the experience of each patient. Ultimately, the healing relationship between a caregiver and a patient is a human one. Our goal has been to bring readers the best wisdom possible from the available technologies. We have no doubt that combining these technologies with human insights and compassionate care will profoundly improve the lives of patients and their families.

REFERENCES

[1] Boissey A, Lee TH. Better communication makes better physicians. NEJM Catal February 20, 2017. http://catalyst.nejm.org/better-communication-makes-better-physicians/.

[2] Kaplan RS, Hass DA, Warsh J. Adding value by talking more. N Engl J Med 2016;375:1918–20.

[3] Hibbard JH, Greene J, et al. What the evidence shows about patient activation: better health outcomes and care experiences; fewer data on costs. Health Aff 2013;32:207–14.

[4] Greene J, Hibbard JH, Sacks R, et al. When patient activation levels change, health outcomes and costs change. Health Aff 2015;34:431–7.

[5] Simmons LA, Wolever RQ, Bechard EM, Snyderman R. Patient engagement as a risk factor in personalized health care: a systematic review of the literature on chronic disease. Genome Med 2014;6:16. http://genomemedicine.com/content/6/2/16.

[6] Chase D. Exclusive book excerpt: engage! Transforming healthcare through digital patient engagement. Forbes July 1, 2013. https://www.forbes.com/sites/davechase/2013/07/01/exclusive-book-excerpt-engage-transforming-healthcare-through-digital-patient-engagement/#5705110e1604.

[7] Schiff GD, Seoane-Vazquez E, Wright A. Incorporating indications into medication ordering—time to enter the age of reason. N Engl J Med 2016;375:306–9.

[8] Institute of Medicine. Crossing the quality chasm: a new health system for the 21st century. Washington (DC): National Academies Press; 2001.

[9] Centers for Medicare and Medicaid Services. Beneficiary engagement and incentives models: shared decision making model. December 8, 2016. https://www.cms.gov/Newsroom/MediaReleaseDatabase/Fact-sheets/2016-Fact-sheets-items/2016-12-08-2.html.

[10] Bottles K, Begoli E. Understanding the pros and cons of big data analytics. Physician Exec 2014;40:6–10.

[11] Halamka J. CareKit as an enabler for patient generated healthcare data. Life Healthc CIO March 23, 2016. http://geekdoctor.blogspot.com/2016/03/carekit-as-enabler-for-patient.html.

[12] Lavallee DC, Chenok KE, Love RM, et al. Incorporating patient-reported outcomes into health care to engage patients and enhance care. Health Aff 2016;35:575–82.

INDEX

Note: "Page numbers followed by "f" indicate figures, "t" indicate tables and "b" indicate boxes."

Made in the USA
Lexington, KY
06 December 2019